CAMBRIDGE STUDIES IN ADVANCED MA

Representations and characters of finite groups

To Marjorie

Contents

Preface

Representation theory and character theory provide major tools for the study of finite groups. Complex representations and their characters were first studied nearly 100 years ago by Frobenius, and his theorem on transitive permutation groups was the first major achievement of the theory; it remains to this day, along with Burnside's $p^a q^b$-theorem, one of the highlights of any first course on representation theory. Indeed, while Burnside's theorem now admits a purely group theoretic proof, Frobenius' theorem remains untouched by non-character-theoretic methods.

Both of these results may be regarded as nonsimplicity criteria. The study and application of character theory, since Brauer proposed a systematic programme to classify the finite simple groups at the Amsterdam International Congress in 1954, cannot be divorced from the classification itself. Although purely group theoretic methods have dominated the major part of that work which took place between 1970 and 1980, the classification could never have been carried out (and, indeed, would have been stillborn) without the early progress made using character theory. The reason for this is quite simple. The goal is always to obtain global information from local information (that is, information about the whole group from information about various subgroups). However, the process is inductive; when a group is in some sense big enough, there is enough interaction between the various subgroups to progress from that interaction by group theoretic means, but in small configurations the information available is so tight (and the situation which occurs in the proof of Frobenius' theorem is a perfect example) that what Brauer referred to as the 'arithmetic' properties of groups, namely their characters, have to be studied. Thus characters have played an indispensable role in the study of certain permutation groups and in the characterisation of groups with Sylow 2-subgroups of small rank (possibly zero), and this will remain necessary in the context of the current revision project.

There have been two major strands to this work; on the one hand, there is Brauer's theory of blocks of characters, arising out of his work on modular representation theory, and, on the other, the theory of exceptional characters developed by Suzuki, Feit and others, as a direct attempt to generalise the original ideas of Frobenius. The aim of this book is to give a comprehensive and self-contained account of the latter, with all the refinements and improvements which are now possible.

The origin of this book lies in graduate lectures that I gave in Oxford over the years 1971–3 and again in 1976–8, together with other courses given since. In the original courses, I covered both the ordinary (complex) theory and modular theory but, with an exception on which I shall comment below, I have concentrated here on those areas of ordinary representation theory and character theory which should occur in any graduate course or which will serve as an introduction for those who wish to study their application to the classification of simple groups, and I have included a substantial amount of material which I did not put into any of the lectures, and some of which is far more recent. In a book of this length, one cannot hope to cover all topics and I have made no such attempt; for example, there is no mention of the character theory of soluble groups or the symmetric groups, nor any of Schur indices beyond the basic Frobenius–Schur index. On the other hand, I have written this book very much for the group theorist so that, for example, the first chapter contains a considerable amount of general material that is relevant to the application of representation theoretic methods to 'internal' group theory, and this book contains sufficient for anyone wishing to go on to study the character theory of the odd order paper of Feit and Thompson, either in the original form or as revised by Sibley.

This book should be accessible both to graduates and to higher level undergraduates, and the range of topics and exercises in the first two chapters which cover basic representation theory and character theory will have such readers in mind, though a judicious choice of material for undergraduate use might be desirable. Only a very basic knowledge of group theory and more general algebra is assumed, though a more sophisticated background would be valuable to see some of the examples and applications in their full perspective; unavoidably, some of the material in these chapters is there for later application. I have, however, tried to prepare the reader for study in areas other than those covered later; there is sufficient background from the representation theory side, for example, to study the Deligne–Lusztig theory of the ordinary characters of groups of Lie type. Also, in the final section of Chapter 2, I have included a brief discussion of the application of the character theoretic methods which are central in this book to the Inverse Galois problem.

After the first two chapters, our treatment becomes more specialised. In Chapter 3, we examine Suzuki's theory of exceptional characters and then, in Chapter 4, Feit's theory of isometries and coherence. Sibley's refinements of Feit's work are discussed here and they are applied to simplify two major applications of character theory; in Section 4.5, we shall establish the nonsimplicity of CN-groups of odd order following Suzuki's treatment of CA-groups of odd order, and, in Section 4.6, we shall give a unified treatment of the reduction theorems for Zassenhaus groups which were originally proved by Feit and by Itô. In Chapter 5, which is self-contained, Brauer's characterisation of characters is discussed; this topic may today be included in any basic course and can be read immediately after Section 2.3, but here is the appropriate place in the context of this book since we wish to emphasise its role in the construction of isometries other than that arising directly from character induction as in the Suzuki theory. This work takes place in the final chapter.

It is in this final chapter that I make the one exception to self-containment. Block theory is needed for Reynolds' work and for the more recent work of Robinson on isometries, for which I have given a unified treatment. I had intended just to state the most basic results required but, as I was writing, I found that all but Brauer's second and third main theorems (which are only stated) can be proved by methods already discussed at length, and I have therefore indulged in a somewhat unusual approach to block theory. That this should prove possible did not come as a surprise. I have long felt that Brauer's method of columns was somehow the dual of Suzuki's method of exceptional characters (and this is implicit in work of Walter and Wong), and I first gave a proof of the nonsimplicity of groups with homocyclic Sylow 2-subgroups of rank 2, replacing the use of Brauer's method of columns by an isometry constructed using Brauer's characterisation of characters, in my lectures; what did surprise me was that I could also construct the principal 2-block for groups with dihedral Sylow 2-subgroups by the same approach. This has allowed me to give in a natural way Suzuki's proof of the Brauer–Suzuki theorem on groups with an ordinary quaternion Sylow 2-subgroup.

The link is Brauer's characterisation of characters, and I have chosen to give the Brauer–Tate proof rather than the shorter proof of Goldschmidt and Isaacs since I believe that it may give greater insight into the relationship between character ring methods and Brauer's methods with modular characters. My one regret is that I cannot justify this comment explicitly; although modular representation theory is very much in vogue today, I would hope that Brauer's second and third main theorems will one day submit to non-modular proofs.

Finally I turn to acknowledgements. I owe my own original interest in character theory to some second year undergraduate lectures 'advertising' algebra which were given in Oxford by Martin Powell some 25 years ago and to the course given by Graham Higman when I was first his research student; Section 3.4, in particular, reflects this upbringing, although Theorem 3.17 turns out to have a new application in my treatment of Zassenhaus groups! Looking back to those formative years, I should also thank Michio Suzuki who sent me a number of his papers, one of which contains the marginal note, 'More work is needed'.

For the metamorphosis of my lectures into this book, I would like to thank several generations of students in Oxford (and one at CalTech) who have unearthed many unclear points (and a few howlers). Jonathan Alperin, Peter Landrock and, in particular, Geoffrey Robinson have read sections of the manuscript at various stages, and I am grateful for all their comments. It would also be appropriate to thank Jack Cowan who introduced me to the Macintosh; without it, I would not have been able so easily to send draft sections around for comment and make almost constant changes to the manuscript. I should also like to thank the Universities of Aarhus, Chicago and Essen for their hospitality and support at various times while this book was being written.

Finally I should like to thank all those at the Cambridge University Press for their assistance – and, in particular, David Tranah for his patience.

MICHAEL COLLINS

Oxford
February, 1989.

1

General representation theory

1 Basic concepts

Let G be an arbitrary finite group[†] and let K be an arbitrary field. Then a *(linear) representation* ρ *of* G *over* K is a homomorphism

$$\rho: G \to \mathrm{GL}(V)$$

where V is a finite dimensional vector space[†] over K and $\mathrm{GL}(V)$ is the group of nonsingular linear transformations of V into itself. If we are already given the vector space V, then we may refer to ρ as a representation of G *on* V. Although we shall normally write mappings on the right, we shall write $\rho(g)$ rather than $g\rho$ since we shall never consider compositions of representations; also, for $v \in V$, we shall write vg for $v \cdot \rho(g)$ if there is no risk of ambiguity.

If the dimension of V is n, we may choose a basis for V and identify V with the space K^n of n-tuples over K; then we may regard ρ as a map from G into $\mathrm{GL}(n, K)$, the group of nonsingular $n \times n$ matrices over K. The precise map so obtained depends on the choice of basis; thus a homomorphism

$$\rho: G \to \mathrm{GL}(n, K)$$

should be called a *matrix* representation. However, in a way which will be made precise shortly, matrix representations obtained by taking different bases are *similar*, and we shall move freely between representations on vector spaces and the corresponding matrix representations.

We shall study representations with two particular purposes in mind. The first is that a representation gives us something concrete, namely a group of linear transformations or matrices, to which the methods of linear algebra may be applied. The second is that by studying the values of the traces of the matrices $\rho(g)$, it may be possible to use the arithmetic properties of the field K to deduce information about an abstract group G. This is known as *character theory*, and much of this book will be devoted to this aspect in the case that K is the field of complex numbers $\mathbb{C}$.

[†] Throughout this book, groups will always be finite, with the obvious exception of groups of linear transformations, and vector spaces will be finite dimensional. However, most of the definitions of this section, although little of the subsequent theory, can be extended without these restrictions.

Such trace values are known as (ordinary) *characters*. However, in this chapter we shall develop the basic representation theory in the first spirit and in a form more general than that needed purely for character theory.

Examples (Groups and fields are arbitrary unless otherwise stated.)

1. Let V be a one-dimensional vector space over K. The map

$$g \to 1_V$$

for all $g \in G$ is the *trivial* representation of G over K.

2. Let G be a group which acts as a group of permutations on a finite set Ω, where $\Omega = \{e_1, \ldots, e_n\}$. Let V be a vector space of dimension n over K with a basis $\{v_1, \ldots, v_n\}$. For $g \in G$, let π_g be the linear transformation on V defined by the action on basis vectors

$$\pi_g: v_i \to v_j \quad \text{if and only if} \quad g: e_i \to e_j.$$

Then the map $\pi: G \to \mathrm{GL}(V)$ defined by $\pi(g) = \pi_g$ for all $g \in G$ is a *permutation* representation of G on V. Notice that the corresponding matrix representation (with respect to the basis $\{v_1, \ldots, v_n\}$) is given by permutation matrices.

3. Take $\Omega = G$ in Example 2 and define a permutation action by the mappings

$$g: x \to xg$$

for all $x, g \in G$. The associated representation is called the *right regular representation* of G.

4. Let N be a normal subgroup of G, and suppose that ρ is a representation of G/N. The mapping

$$\hat{\rho}: g \to \rho(gN)$$

for all $g \in G$ defines a representation on G. This representation is called the *inflation* of ρ.

Conversely, if σ is a representation of G such that N lies in the kernel of σ, then the mapping

$$\tilde{\sigma}: gN \to \sigma(g)$$

defines a representation of G/N.

5. If K is regarded as a one-dimensional vector space over itself, then multiplication acts as a linear transformation. Thus any homomorphism from a group G into the multiplicative group of K may be viewed as a representation. In particular, if ρ is a representation of G over K, then

the mapping

$$g \to \det(\rho(g))$$

for all $g \in G$ is a one-dimensional representation.

6. Let G be a cyclic group of order n and let g be a generator of G. Let ω be an nth root of unity in K (not necessarily primitive). Then the mapping

$$\rho_\omega: g^i \to \omega^i$$

defines a representation of G. Conversely, every one-dimensional representation of G is similar to a representation of this form.

7. The alternating groups A_4 and A_5 and the symmetric group S_4 are isomorphic, respectively, to the rotation groups of the regular tetrahedron, icosahedron and cube. By taking an orthonormal basis for $\mathbb{R}^3$, these isomorphisms lead to natural representations of the three groups by real *orthogonal* 3×3 matrices.

We shall now introduce some basic terminology. Let G and K be, as before, arbitrary and let ρ be a representation of G on a vector space V over K. The dimension $\dim_K(V)$ is called the *degree* of the representation ρ and will be denoted by $\deg \rho$. If the kernel, $\ker \rho$, of ρ is trivial, then ρ is *faithful*. If U is a subspace of V which is invariant under $\rho(g)$ for all $g \in G$, then U *admits* G, or is *G-invariant*. If $V \neq 0$ and the only G-invariant subspaces of V are 0 and V itself, then ρ is *irreducible*; otherwise ρ is *reducible*. If V can be written as the direct sum of two nonzero G-invariant subspaces, then ρ is *decomposable*; otherwise ρ is *indecomposable*.

It follows, trivially, that an irreducible representation is indecomposable. The converse is true provided that the characteristic of K does not divide[†] the order of G as we shall see in Section 3, but this need not be so in general. For example, a two-dimensional representation of the additive group of $\mathbb{Z}_p$ over $\mathbb{Z}_p$ is given by

$$t \to \begin{pmatrix} 1 & t \\ 0 & 1 \end{pmatrix},$$

and this is indecomposable but not irreducible: the subspace spanned by the second basis vector is invariant, but not complemented.

Suppose that ρ_1 and ρ_2 are two representations of G over K on vector spaces V_1 and V_2 respectively. Then ρ_1 and ρ_2 are said to be *equivalent* if there exists an isomorphism $\sigma: V_1 \to V_2$ such that

$$\rho_2(g) = \sigma^{-1}\rho_1(g)\sigma \quad \text{for all } g \in G;$$

we shall write $\rho_1 \sim \rho_2$, or $\rho_1 \sim_K \rho_2$ if we wish to emphasise the field K.

[†]This will always be understood to include the case that char $K = 0$.

Visibly, this defines an equivalence relation on the representations of a group (over a fixed field), and by a set of *distinct* representations of a group, we shall always mean a collection of inequivalent representations.

If $V_1 = V_2$, then this definition can be applied to two different representations of G on V_1; also, if ρ is a representation of G on V and ρ_1 and ρ_2 are the associated *matrix* representations with respect to different bases of V, then immediately ρ_1 and ρ_2 are equivalent. In this case, there exists a nonsingular matrix X such that

$$\rho_2(g) = X^{-1}\rho_1(g)X$$

for all $g \in G$, and we say that ρ_1 and ρ_2 are *similar*.

Exercises

1. Show that the derived group G' of a group G lies in the kernel of any representation of G of degree 1. Deduce that, if $\rho: G \to \mathrm{GL}(n, K)$ is a matrix representation of G, then $\rho(g) \in \mathrm{SL}(n, K)$ whenever $g \in G'$.
2. Let ρ_1 and ρ_2 be equivalent representations of a group G. Show that, whenever $g \in G$, the linear transformations $\rho_1(g)$ and $\rho_2(g)$ have the same minimal and characteristic polynomials. If g is an element of order n, show that the minimal polynomial of $\rho_1(g)$ divides $x^n - 1$.
3. Let ρ be a representation of a group G over an algebraically closed field K whose characteristic does not divide $|G|$. If g is a fixed element of G, show that there exists a basis with respect to which $\rho(g)$ has a diagonal matrix.
4. Let G be a finite abelian group and let K be an algebraically closed field of characteristic not dividing $|G|$. If G has a representation on a vector space V over K, show that there exists a basis for V with respect to which every element of G is represented by a diagonal matrix. Deduce that every irreducible representation of G over K has degree 1.
5. Let G and K be as in Exercise 4 and regard the irreducible representations as maps from G to K. Suppose that G has a decomposition as the direct product of cyclic subgroups generated by elements $g_1, \ldots, g_r$. Show that an irreducible representation ρ of G is determined by its values on the elements $g_1, \ldots, g_r$ alone. Deduce that the number of distinct irreducible representations of G over K is $|G|$.

   Show that the set of distinct irreducible representations forms an abelian group G^* under composition defined by

   $$(\rho_1 \cdot \rho_2)(g) = \rho_1(g) \cdot \rho_2(g) \quad \text{for all } g \in G,$$

   and that G^* is isomorphic (as an abstract group) to G.

6. Show that an irreducible representation of a cyclic group G of prime order p over a field of characteristic p is trivial. By considering the possible Jordan canonical forms for a linear transformation of order p, determine a complete set of inequivalent indecomposable representations of G over a field of characteristic p.
7. Determine the irreducible representations of an arbitrary p-group over a finite field K of characteristic ρ.
   [Hint. Show that the subgroup of upper triangular matrices in $\mathrm{GL}(n, K)$ is a Sylow subgroup, where n is the degree, and apply Sylow's theorem.]
8. Let G be the dihedral group D_{2n} of order $2n$, the group of symmetries of a regular n-gon. Then G has a presentation
$$G = \langle x, y \,|\, x^n = y^2 = 1, y^{-1}xy = x^{-1} \rangle.$$
   Suppose that n is odd. Show that $G' = \langle x \rangle$, and hence determine the (two) one-dimensional representations of G over $\mathbb{C}$.
   Use Exercises 1 and 4 to show that, if $\rho: G \to \mathrm{GL}(2, \mathbb{C})$ is a two-dimensional irreducible complex representation of G, then $\rho \sim \rho_1$ where
$$\rho_1(x) = \begin{pmatrix} \omega & 0 \\ 0 & \omega^{-1} \end{pmatrix}$$
   and ω is a nonidentity nth root of unity. Determine which matrices of order 2 can invert $\rho_1(x)$, and hence show that $\rho_1 \sim \rho_2$ where
$$\rho_1(x) = \rho_2(x) \quad \text{and} \quad \rho_2(y) = \begin{pmatrix} 0 & 1 \\ 1 & 0 \end{pmatrix}.$$
   Deduce that the number of inequivalent irreducible complex representations of G of degree 2 is $\frac{1}{2}(n-1)$.
   [Notice that $\frac{1}{2}(n-1)\cdot 2^2 + 2\cdot 1^2 = 2n$: see Exercise 17 of Section 2 and also Corollary 20 (iii).]
9. Carry out the corresponding analysis to Exercise 8 when n is even.
   [Note that, in this case, $G' = \langle x^2 \rangle$.]
10. Let G be the generalised quaternion group of order 2^{n+1} ($n \geqslant 2$) which has a presentation
$$g = \langle x, y \,|\, x^{2^n} = 1, y^2 = x^{2^{n-1}}, y^{-1}xy = x^{-1} \rangle.$$
    Show that there is a complex representation $\rho: G \to \mathrm{GL}(2, \mathbb{C})$ for which
$$\rho(x) = \begin{pmatrix} \omega & 0 \\ 0 & \omega^{-1} \end{pmatrix} \quad \text{and} \quad \rho(y) = \begin{pmatrix} 0 & -1 \\ 1 & 0 \end{pmatrix}$$
    where ω is a primitive 2^nth root of unity and that ρ is faithful and

irreducible. Show also that every element of G is represented by a matrix in $SL(2, \mathbb{C})$, and deduce directly that G contains a unique involution (element of order 2).

11. In Example 7, use geometrical considerations to show that the representations defined there are irreducible.
12. Let G be a group, and define an action of G as a group of permutations on itself by $g: x \to g^{-1}x$. Show that this gives rise to a representation over any field, called the *left regular representation.*
13. Show that, if ρ and σ are similar matrix representations of a group G over a field K, then $\operatorname{tr}(\rho(g)) = \operatorname{tr}(\sigma(g))$ for all $g \in G$.

    Show also that $\operatorname{tr}(\rho(g)) = \operatorname{tr}(\rho(h))$ whenever g and h are conjugate elements in G. (This says that we have defined a *class function* on G.) [tr denotes the *trace* of matrix: if $A = (a_{ij})$, then $\operatorname{tr} A = \sum_i a_{ii}$.]
14. For each of the groups A_4, A_5 and S_4, determine the value of $\operatorname{tr}(\rho(g))$ for a representative of each conjugacy class, where ρ is the three-dimensional real representation defined for each of the three groups in Example 7.

2 Group rings, algebras and modules

Let G be a finite group and let ρ be a representation of G on a vector space V over a field K. Then the K-linear combinations of the linear transformations $\rho(g)$ for $g \in G$ form a subring of the full ring $\mathscr{L}(V)$ of linear transformations of V. The vector space V can be given the structure of a right module over this subring. We shall formalise this, but make our first definition more general.

Let G be a group and let R be a commutative ring with identity. Then the *group ring* RG consists of the set of all formal sums

$$\sum_{g \in G} a_g g \quad (a_g \in R)$$

together with the binary operations

$$\sum_{g \in G} a_g g + \sum_{g \in G} b_g g = \sum_{g \in G} (a_g + b_g) g \quad (a_g, b_g \in R)$$

$$\left(\sum_{g \in G} a_g g \right) \cdot \left(\sum_{h \in G} b_h h \right) = \sum_{g \in G} \left(\sum_{h \in G} a_{gh^{-1}} b_h \right) g$$

$$= \sum_{g, h \in G} (a_g b_h)(gh)$$

where gh is the group product in G. It is a straightforward calculation to verfiy that RG is an associative ring with identity. If R is a field K, then KG has the structure of a vector space over K as well as that of ring. So in this case KG is a K-algebra of finite dimension $|G|$, called the *group algebra* of G over K.

Let $\rho: G \to \mathrm{GL}(V)$ be a representation of G on a vector space V over the field K. Then we can extend ρ by linearity to a K-algebra homomorphism

$$\rho: KG \to \mathrm{Hom}_K(V, V).$$

The extension ρ is a representation of the algebra KG. This gives V the structure of a unitary right KG-module under the operation

$$v(\sum a_g g) = \sum a_g(v \cdot \rho(g)),$$

called the *representation module*. Since $\dim_K(V)$ is finite, V certainly satisfies both chain conditions as a KG-module; the composition factors (or the representations that they *afford*–see Propositions 1 and 2 below) are called the *irreducible constituents* of ρ. Conversely, given a unitary right KG-module M which is finite dimensional as a vector space over K, we may obtain a representation σ of G defined by $m \cdot \sigma(g) = mg$. In our approach, we shall switch freely between modules and representations.

These considerations extend naturally to the representations (over K) of an arbitrary K-algebra $A^{\dagger}$. Conversely, given an A-module $M^{\dagger}$, we may recover a representation of A. It is easy to see that the definitions of Section 1 extend to representations of K-algebras (over K) and then to verify the following.

Proposition 1. *Let G be a group and K a field. Then a representation of G over K is irreducible or indecomposable if and only if the same is true of the corresponding representation of KG. Two representations of G are equivalent if and only if the same is true for the corresponding representations of KG.*

Proposition 2. *Let A be an algebra over a field K and let M be an A-module which affords a representation ρ of A. Then*

(i) *M is irreducible as an A-module if and only if ρ is an irreducible representation,*
(ii) *M is indecomposable as an A-module if and only if ρ is an indecomposable representation, and*
(iii) *if $M = M_1 \oplus \cdots \oplus M_n$ is a direct sum decomposition of M into a sum of indecomposable submodules, then n and the isomorphism classes of $M_1, \ldots, M_n$ are uniquely determined.*

† Algebras will always be associative with identity and finite dimensional. Modules will be unitary right modules unless explicitly stated otherwise, and modules over algebras will be finite dimensional over the underlying field.

Furthermore, if N is an A-module affording a representation σ, then ρ and σ are equivalent if and only if M and N are isomorphic as A-modules.

Proof. All but (iii) are immediate from the definitions. We note that M always has such a decomposition since it satisfies both chain conditions on submodules: then the Krull–Schmidt theorem holds.

We shall now consider some constructions of representations that will be used later, leaving the necessary verifications as exercises.

Examples

1. If $\rho: G \to \mathrm{GL}(V)$ is a representation of G and H is a subgroup of G, then the *restriction* $\rho|_H: H \to \mathrm{GL}(V)$ is a representation of H. Since $KH \subseteq KG$, a KG-module M has the structure of a KH-module, which we denote by M_H.

2. If M_1 and M_2 are KG-modules affording representations ρ_1 and ρ_2 respectively, then their direct sum $M_1 \oplus M_2$ affords the *sum* $\rho_1 + \rho_2$.

3. Let M be a KG-module affording a representation ρ, and let M^* be the dual of M as a vector space over K. Then M^* may be given the structure of a KG-module as follows. For $m^* \in M^*$ and $g \in G$, define

$$m^*g = \rho(g^{-1}) \circ m^*;$$

then $m^*g \in M^*$ and we put

$$m^*(\sum a_g g) = \sum a_g(m^*g).$$

M^* affords the *contragredient representation* ρ^*: as matrix representations with respect to dual bases, $\rho^*(g)$ will be the transpose inverse of $\rho(g)$ for each $g \in G$.

4. Let $\rho: G \to \mathrm{GL}(n, K)$ be a matrix representation, and suppose that α is an automorphism of K. If $\rho(g) = (a_{ij}(g))$, then a representation ρ^α can be defined by putting $\rho^\alpha(g) = ((a_{ij}(g))^\alpha)$. An example of particular importance occurs when $K = \mathbb{C}$ and α is complex conjugation.

5. Let ρ be a representation of a group G and let θ be an automorphism of G. Then the map $g \to \rho(g\theta)$ defines a representation of G. (This is just the composite of θ and ρ.) Notice that this action of θ on sets of representations preserves equivalence.

6. Let G and H be groups and let M and N be KG- and KH-modules respectively affording representations ρ and σ of G and H. Let $M \otimes N$ be the tensor product of M and N as vector spaces. Then $M \otimes N$ can be

given the structure of a $K(G \times H)$-module by defining

$$(m \otimes n)(g, h) = mg \otimes nh$$

for decomposable tensors and extending by linearity. $M \otimes N$ affords the *tensor product* representation $\rho \otimes \sigma$ of $G \times H$.

7. Take $G = H$ in Example 6 above and restrict $\rho \otimes \sigma$ to the diagonal subgroup of $G \times G$. Then $M \otimes N$ affords a representation $\rho \otimes \sigma$ of G, also called the tensor product, under the operation

$$(m \otimes n)g = mg \otimes ng.$$

8. Let M be an A-module, where A is a K-algebra, and suppose that L is an extension of K. Put $A^L = L \otimes_K A$ and $M^L = L \otimes_K M$. Then A^L may be regarded as an L-algebra and M^L as an A^L-module under the operation

$$(l \otimes m)(l' \otimes a) = ll' \otimes ma.$$

Notice that $\dim_L(M^L) = \dim_K(M)$. In the case of group algebras, there is a natural isomorphism between $(KG)^L$ and LG. Regarding M^L as an LG-module and letting M and M^L afford representations ρ and ρ^L of G respectively, the corresponding matrix representations of G will be identical with respect to bases of the form $\{m_i\}$ and $\{1 \otimes m_i\}$. Thus, by abuse of terminology, $\rho \sim \rho^L$ and characteristic and minimal polynomials are unaltered by field extension. Usually we shall follow this abuse and refer to *extending the ground field*, rather than actually perform the tensor product construction since this is what happens in reality when considering representations of groups and embedding $\mathrm{GL}(n, K)$ in $\mathrm{GL}(n, L)$.

If ρ is an irreducible representation of G over K, then ρ is *absolutely irreducible* if ρ^L is irreducible whenever L is an extension of K, and K is a *splitting field* for G if every irreducible representation of G over K is absolutely irreducible. As will be seen in Exercise 12, any extension of a splitting field is a splitting field: no 'new' irreducible representations occur in a larger field. These definitions extend to algebras and modules provided the precise formulation is taken.

We shall return to the study of absolute irreducibility in Section 4.

Exercises

1. If G is a group and N is a normal elementary abelian p-subgroup of G, show that N may be given the structure of a $\mathbb{Z}_p(G/N)$-module by defining

$$(n)gN = g^{-1}ng$$

for all $n \in N$ and $g \in G$. What is the kernel of the corresponding representation of G?

2. Show that, if G is a group and M and N are normal subgroups of G with $N \subset M$ and M/N an elementary abelian p-group, then M/N may be given the structure of a $\mathbb{Z}_pG$-module.
3. Show that if H is a subgroup of a group G and M and M' are KG-modules, then $(M \oplus M')_H \cong M_H \oplus M'_H$ and $(M \otimes M')_H \cong M_H \otimes M'_H$. Show also that $(M^*)_H \cong (M_H)^*$.
4. Show that the module given by a permutation representation of a group always contains a submodule affording the trivial representation.
5. Show that a group ring KG, viewed as a module over itself, has a unique submodule affording the trivial representation, spanned (as a vector space) by $\sum g$.
6. Show that a group ring KG affords the right regular representation of G when viewed as a right module over itself. What can be said of KG as a *left* module?
7. Let H be a normal subgroup of a group G and let K be a field. Show that KH can be given the structure of a KG-module by defining
$$h \cdot g = g^{-1}hg$$
and extending linearly.
8. Let A be a K-algebra. Show that, as a module over itself, the submodules of A are precisely the right ideals of A.
9. Let A be a K-algebra and let M be an irreducible A-module. Show that A has a maximal right ideal B such that $A/B \cong M$ as an A-module. Deduce that A has only finitely many inequivalent irreducible representations.
10. Let A be a K-algebra. Show that the intersection of all maximal right ideals forms a two-sided ideal $J(A)$, called the *radical* of A, and that
$$J(A) = \{a \in A \mid Ma = 0 \text{ for every irreducible } A\text{-module } M\}.$$
Show also that $J(A/J(A)) = 0$.
11. Let M be an irreducible KG-module where char K divides $|G|$. For $m \in M$, show that, if $m(\sum g) \neq 0$, then $m(\sum g)$ spans a subspace on which G acts trivially. Deduce that, in fact, $m(\sum g) = 0$ and hence that $J(KG) \neq 0$.
12. Let A be a K-algebra and let M be an A-module. Suppose that L is an extension of K. Show that

    (i) if N is an A-submodule of M, then N^L may be regarded as an A^L-submodule of M^L and $M^L/N^L \cong (M/N)^L$,
    (ii) if E is an extension of L, then $M^E \cong (M^L)^E$, and
    (iii) if M is absolutely irreducible, then so is M^L.

Suppose that K is a splitting field for A and that $\{M_1, \ldots, M_r\}$ is a complete set of nonisomorphic irreducible A-modules. If $\tilde{M}$ is an irreducible A^L-module, show that $\tilde{M} \cong M_i^L$ for some i and deduce that L is also a splitting field for A.
[Note. This is also a consequence of theorems that will be proved later, but it is instructive to find an elementary proof. It will also follow from the structure theorems that no two of $M_1^L, \ldots, M_r^L$ are isomorphic. See Theorem 13 and Exercise 8 of Section 4.]

13. Let ρ be a representation of a group G over a field K, where char K does not divide $|G|$. Show that there exists a finite extension L of K such that ρ is similar over L to a representation in which any specified element g may be represented by a diagonal matrix. What restriction need be imposed to obtain a like conclusion if char K does divide $|G|$?
14. Let G be an abelian group of exponent n. Show that if K is a field whose characteristic does not divide $|G|$ and K contains primitive nth roots of unity, then K is a splitting field for G. Show that if the second condition is not satisfied, then K is not a splitting field.
[See Exercises 4 and 5 of Section 1. The *exponent* of a group is the least common multiple of the orders of its elements.]
15. Let G a group and H a subgroup, and let K be a field. Show that, if $\{g_1, \ldots, g_n\}$ is a set of coset representatives for H in G, then the subspace of KG spanned by the coset sums

$$\sum_{h \in H} hg_i, \quad i = 1, \ldots, n,$$

forms a submodule.
16. Determine the irreducible constituents of the right regular representation of the symmetric group S_3 over each of the fields $\mathbb{C}, \mathbb{Q}, \mathbb{Z}_2$, and $\mathbb{Z}_3$ and deduce that each is a splitting field for S_3.
17. Let $G = D_{2n}$ and suppose that n is odd. Using the notation of Exercise 8 of Section 1, show that, if ω is a nonidentity nth root of unity, the complex group algebra $\mathbb{C}G$ has a two-dimensional subspace on which x acts as multiplication by ω in the right regular representation. By considering the action of y on the four-dimensional subspace spanned by the eigenspaces of x corresponding to eigenvalues ω and ω^{-1}, show that $\mathbb{C}G$ has at least two composition factors affording the representation ρ_2. Deduce, by considering the sum of their dimensions, that $\mathbb{C}G$ has exactly one composition factor affording each one-dimensional representation and exactly two composition factors affording each two-dimensional irreducible representation.

[Note. Compare this with the previous exercise. Exercise 17 now shows that every irreducible representation of G over $\mathbb{C}$ has been described. See also Corollary 20(ii).]

18. Obtain the analogue of the previous exercise with n even.
19. Show that A_5 has two inequivalent real representations of degree 3.
    [Hint. Consider the effect of conjugation by a 4-cycle in S_5 on a 5-cycle that it normalises, and trace values. See Example 7 and Exercise 14 of Section 1.]
20. Let M and N be A-modules for a K-algebra A. Show that $\mathrm{Hom}_K(M, N)$ can be given the structure of an A-module by defining
    $$m(fa) = (mf)a$$
    for $m \in M, f \in \mathrm{Hom}_K(M, N)$ and $a \in A$, where $\mathrm{Hom}_K(M, N)$ denotes the vector space of linear transformations from M to N.
21. Let M and N be KG-modules. Show that $\mathrm{Hom}_K(M, N)$ can be given the structure of a KG-module by defining a map f^g by
    $$mf^g = (mg^{-1}f)g$$
    for all $m \in M$ and extending by linearity. Show that the space of fixed points of the action of G on $\mathrm{Hom}_K(M, N)$ is $\mathrm{Hom}_{KG}(M, N)$, the set of homomorphisms from M to N as KG-modules.
22. Let K be an arbitrary field. Show that any $n \times n$ matrix over K can be expressed as a finite K-linear combination of nonsingular matrices over K.
    [In this sense, there exists a homomorphism from the group ring of $\mathrm{GL}(n, K)$ over K, taking only finitely many nonzero terms, onto the full matrix ring $\mathcal{M}_n(K)$.]
23. Let M be an A-module and define the *symmetric* and *antisymmetric* subspaces of $M \otimes M$ as the subspaces M_S and M_A spanned by the sets $\{m \otimes m \mid m \in M\}$ and $\{(m_1 \otimes m_2 - m_2 \otimes m_1) \mid m_1, m_2 \in M\}$ respectively. Show that M_S and M_A are submodules and that $M \otimes M = M_S \oplus M_A$ if char $K \neq 2$.

3 Complete reducibility

Let K be an arbitrary field and let A be a K-algebra. An A-module M is said to be *completely reducible* if the conditions of the following proposition hold. The representation of A afforded by M will also be said to be completely reducible.

Proposition 3. *Let M be an A-module. Then the following three conditions are equivalent.*

(i) *M is a direct sum of irreducible submodules.*
(ii) *M is a sum of irreducible submodules.*
(iii) *Every submodule of M is a direct summand.*

Proof. Clearly (i) implies (ii). If (ii) holds, we establish (iii) by induction on codimension. Let N be a submodule of M. If $N = M$, there is nothing to prove; otherwise, by (ii), there exists an irreducible submodule L of M such that $L \cap N = 0$. Then $N + L = N \oplus L$ and, by induction, there exists a submodule L' of M such that

$$M = (N \oplus L) \oplus L' = N \oplus (L \oplus L').$$

If (iii) holds, let $\{M_1, \ldots, M_n\}$ be a collection of irreducible submodules of M which generate their direct sum M'. If $M' \neq M$, then we can choose a submodule M'' of M and an irreducible submodule M_{n+1} of M'' such that $M = M' \oplus M''$ and

$$(M_1 \oplus \cdots \oplus M_n) + M_{n+1} = M_1 \oplus \cdots \oplus M_n \oplus M_{n+1} \supset M_1 \oplus \cdots \oplus M_n.$$

Since M is finite dimensional over K, it follows that M is a direct sum of irreducible submodules.

For the remainder of this section, we shall restrict our attention to representations of group algebras. The next result will provide a fundamental dichotomy according to the characteristic of the underlying field.

Theorem 4 (Maschke). *Let G be a finite group and let K be a field whose characteristic does not divide $|G|$. Then every KG-module is completely reducible.*

Proof. Suppose that M is a reducible KG-module and let U be a proper, nonzero submodule. Then we must show that U is a direct summand of M. Let V be a complement to U in M, viewed only as a vector space, and let $\theta: M \to V$ be the corresponding projection. We apply an averaging process to find a KG-invariant complement to U.

Define a map $\varphi: M \to M$ by the formula

$$m\varphi = |G|^{-1} \sum_{g \in G} ((mg)\theta)g^{-1};$$

clearly φ is K-linear. Put $W = M\varphi$. Then $M = U + W$ since, if $m \in M$,

$$m - m\varphi = |G|^{-1} \sum_{g \in G} (mg - (mg)\theta)g^{-1} \in U.$$

Also, if $h \in G$,

$$(m\varphi)h = |G|^{-1} \sum_{g \in G} ((mg)\theta)g^{-1}h$$

$$= |G|^{-1} \sum_{g \in G} ((mh(g^{-1}h)^{-1})\theta)g^{-1}h$$

$$= |G|^{-1} \sum_{x \in G} ((mhx)\theta)x^{-1}$$

$$= (mh)\varphi$$

where we put $x = (g^{-1}h)^{-1}$. So W is G-invariant. Now $\varphi|_U = 0$ since $\theta|_U = 0$. Hence, as φ is K-linear,

$$\dim_K(U) + \dim_K(W) \leqslant \dim_K(M),$$

and so $M = U \oplus W$ as a KG-module.

If the characteristic of K does divide $|G|$, then KG is not completely reducible; we shall leave a proof to the exercises below. After the general considerations of this chapter, we shall be restricting our attention to the *ordinary* representation theory of finite groups–namely, that over the complex field, or at least a splitting field of characteristic zero. Here, the complete reducibility of the group algebra will be crucial in developing the basic formulae on which character theory will depend. Analogous results for other fields whose characteristic does not divide the group order hold, but are of limited interest. If the characteristic of the field does divide the group order, then one speaks of *modular* representation theory as was first fully explored by Brauer after earlier work by Dickson. In particular, Brauer developed an extensive theory on the connection between ordinary and modular representations. (Strictly speaking, modular representation theory refers to the situation for fields of nonzero characteristic but, as remarked above, in the coprime characteristic case, this is the same as the ordinary theory.)

Exercises

1. Show that every submodule and factor module of a completely reducible module is completely reducible. Show also that the intersection of all maximal submodules of a completely reducible module is the zero submodule. Deduce that if K is a field whose characteristic divides the order of a group G, then KG is not completely reducible as a KG-module. (See Exercise 11 of Section 2). [Note. This shows that the conclusion of Maschke's theorem would be false if the hypothesis about the field characteristic were omitted. See also Exercise 6.]

2. Show that any KG-module has a unique submodule maximal with respect to admitting G trivially.
3. Let A be a K-algebra and let M be an A-module. Define the *socle* $\mathrm{soc}(M)$ to be the sum of all the irreducible submodules of M and the *radical* $\mathrm{rad}(M)$ to be the intersection of all maximal submodules of M. Prove that $\mathrm{soc}(M)$ and $M/\mathrm{rad}(M)$ are completely reducible.
4. Let A be a K-algebra and let B be a right ideal of A. Prove that A/B is completely reducible as an A-module if and only if $B \supseteq J(A)$. [See Exercise 10 of Section 2.]
5. Show that the set of elements $\sum a_g g$ in a group algebra KG for which $\sum a_g = 0$ forms an ideal $A(KG)$, called the *augmentation ideal*. Show also that $KG/A(KG)$ is isomorphic to the trivial KG-module.
6. Show that, if $\mathrm{char}\, K$ divides $|G|$, then $A(KG)$ contains the unique trivial submodule of KG. Deduce that, in this case, KG has at least two trivial constituents as a KG-module. [Note. KG contains a unique trivial submodule by Exercise 5 of Section 2.]
7. Verify directly that the group algebra of the symmetric group S_3 is completely reducible for the fields $\mathbb{C}$ and $\mathbb{Q}$, but not $\mathbb{Z}_2$ or $\mathbb{Z}_3$. [Use the computations from Exercise 16 of Section 2.]
8. Let A be a K-algebra and let M be an A-module. Identify $\mathrm{Hom}_K(M, M)$ with the full matrix algebra $\mathcal{M}_n(K)$ where $n = \dim_K(M)$ and let A act on $\mathcal{M}_n(K)$ via the formula $m(fa) = (mf)a$. By considering the submodules consisting of matrices with nonzero entries only in a single row, show that $\mathrm{Hom}_K(M, M)$ is isomorphic as an A-module to a direct sum of n copies of M. Deduce that, if M is irreducible, then $\mathrm{Hom}_K(M, M)$ is completely reducible.
9. Let R be an arbitrary ring. Show that the equivalence of the three statements in Proposition 3 holds for R-modules, with no requirement of a finiteness condition (in that case, finite dimensionality).
10. Let P be an elementary abelian p-group and let H be a p'-group which acts on P. By viewing P as a $\mathbb{Z}_p H$-module, prove that $P = C_P(H) \times [P, H]$. [A *p'-group* is a group of order not divisible by a prime p.]

4 Absolute irreducibility and the realisation of representations

We begin by studying an irreducible module M for a K-algebra A and its endomorphism ring $\mathrm{Hom}_A(M, M)$. Our initial goal will be to determine a criterion for M to be absolutely irreducible, namely that only the scalar transformations commute with the action of A. Then we will specialise to the study of group rings and their splitting fields in characteristic 0. The approach that we take is via the double centraliser lemma (Theorem 7); this will in fact yield the full structure of the irreducible representations as a consequence, and we shall obtain this in the next section.

In the following result, we shall identify the field K with the space of scalar transformations: we shall often do this without further comment. The first part is known as *Schur's lemma.*

Theorem 5. $\mathrm{Hom}_A(M, M)$ *is a division ring. If* K *is algebraically closed, then* $\mathrm{Hom}_A(M, M) = K$.

Proof. Since M is irreducible, every endomorphism is either zero or bijective and, in the latter case, it is a trivial verification to show that inverses are also endomorphisms.

Let $\varphi \in \mathrm{Hom}_A(M, M) - \{0\}$. Then φ has a minimal polynomial $f(x)$ since M is finite dimensional over K. If K is algebraically closed, we may write

$$f(x) = \prod_{i=1}^{m} (x - a_i)$$

for some $a_1, \ldots, a_m \in K$. As $(\varphi - a_i 1)$ is either zero or invertible for each i, it follows that $\varphi = a_i 1$ for some i.

We now investigate the structure of M further. Let A_M denote the image of A in $\mathrm{Hom}_K(M, M)$. Then M is irreducible as an A_M-module. Since $\mathrm{Hom}_K(M, M)$ is completely reducible as an A-module, so is A_M, and A_M is isomorphic to a direct sum of copies of M. (See Exercise 8 of Section 3.) Now suppose that I is a proper two-sided ideal of A_M. Then A_M contains a minimal right ideal M_0, isomorphic to M as an A_M-module, with $I \cap M_0 = 0$. So $M_0 I \subseteq M_0 \cap I = 0$, and hence $I = 0$ since A_M acts faithfully on M. Thus we have shown the following.

Lemma 6. A_M *is a simple* K*-algebra.*

Now let $D = \mathrm{Hom}_A(M, M)$. Then M has the structure of a right D-module. Identifying K with the scalar transformations, we see that K lies in the centre of D and hence that $\mathrm{Hom}_D(M, M) \subseteq \mathrm{Hom}_K(M, M)$.

Theorem 7 (Double centraliser lemma). $\mathrm{Hom}_D(M, M) = A_M$.

Proof. Without loss, we may suppose that $A = A_M$ and that $M \subseteq A$. From the definition of D we know that $A \subseteq \mathrm{Hom}_D(M, M)$, so we need only establish the reverse inclusion.

Let $\theta \in \mathrm{Hom}_D(M, M)$. For each $m \in M$, define a map $\theta_m: M \to A$ by $x\theta_m = mx$. As M is a right ideal of A, in fact $\theta_m: M \to M$; then $\theta_m \in D$ since, whenever $a \in A$ and $x \in M$,

$$(xa)\theta_m = m(xa) = (mx)a = (x\theta_m)a.$$

Thus, if $m, n \in M$,

$$(mn)\theta = (n\theta_m)\theta = (n\theta)\theta_m = m(n\theta). \tag{4.1}$$

Let $n \in M - \{0\}$. Since A is simple and has an identity, $AnA = A$ so that we can write

$$1 = \sum_i a_i n b_i \tag{4.2}$$

for suitable elements $\{a_i, b_i\}$ in A. So, whenever $m \in M$, we obtain, using (4.1) and (4.2), the formula

$$m\theta = \sum((ma_i)(nb_i))\theta = \sum(ma_i)((nb_i)\theta) = m\sum a_i((nb_i)\theta).$$

Thus θ acts via right multiplication by an element of A; that is, $\theta \in A$.

If we identify $\mathrm{Hom}_K(M, M)$ with the full matrix algebra $\mathscr{M}_n(K)$ where $n = \dim_K(M)$ and let L be an extension of K, we can see that $\mathrm{Hom}_L(M^L, M^L)$ may be identified with $L \otimes_K \mathrm{Hom}_K(M, M)$. So we obtain the following.

Corollary 8. *Suppose that* $\mathrm{Hom}_A(M, M) = K$. *Then* $A_M = \mathrm{Hom}_K(M, M)$. *If* L *is an extension field of* K, *then* $(A_M)^L = \mathrm{Hom}_L(M^L, M^L)$ *and* $\mathrm{Hom}_{A^L}(M^L, M^L) = L$.

We are now in a position to establish a criterion for absolute irreducibility. This involves only the module M, and does not require a consideration of extension fields.

Theorem 9. *M affords an absolutely irreducible representation of A if and only if* $\mathrm{Hom}_A(M, M) = K$.

Proof. Suppose that $\mathrm{Hom}_A(M, M) = K$. If L is an extension field of K, then in a natural sense $\mathrm{Hom}_L(M^L, M^L) \supseteq (A^L)_{M^L} \supseteq (A_M)^L$ so that, by Corollary 8,

$$(A^L)_{M^L} = \mathrm{Hom}_L(M^L, M^L).$$

So certainly M^L is irreducible as an A^L-module.

Conversely, suppose that $\mathrm{Hom}_A(M, M) \neq K$. Let $\varphi \in \mathrm{Hom}_A(M, M) - K$, and let L be an extension field of K which contains a root λ of the characteristic polynomial $c(x)$ of φ. By assumption, $c(x) \neq (x-\lambda)^n$ where $n = \dim M$, and hence

$$1 \otimes \varphi = \varphi^L \neq \lambda 1.$$

Since $1 \otimes \varphi$ has an eigenvector in M^L, it follows that $(1 \otimes \varphi - \lambda \otimes 1)$ is nonzero and not a unit in $\mathrm{Hom}_{A^L}(M^L, M^L)$. Thus $\mathrm{Hom}_{A^L}(M^L, M^L)$ is not a division ring and so, by Schur's lemma, M^L is reducible.

Combining this result with Theorem 5, we deduce the following.

Corollary 10. *If K is algebraically closed, then K is a splitting field for A.*

Thus the existence of splitting fields has been established. In particular, the complex numbers $\mathbb{C}$ and the algebraic closure of the rationals, which we will denote by $\mathbb{A}$, are splitting fields for an arbitrary finite group. We shall improve on this later, but first introduce the idea of an intertwining matrix.

Let M and N be arbitrary A-modules affording matrix representations ρ and σ of A of degree m and n respectively. We shall say that the $m \times n$ matrix X with coefficients in K *intertwines* the representations ρ and σ if, for all $a \in A$,

$$\rho(a)X = X\sigma(a), \tag{4.3}$$

or that X is an *intertwining matrix* for the two representations. We observe that ρ and σ will be similar if and only if they have the same degree and there exists a nonsingular intertwining matrix. By considering the action of either side of (4.3) on spaces of row vectors, the following proposition is readily verified.

Proposition 11. *Let M and N be A-modules affording matrix representations ρ and σ of A. Then the set of intertwining matrices forms a vector space isomorphic to* $\mathrm{Hom}_A(M, N)$.

We may now define the *intertwining number* $i(M, N)$ for the modules M and N by

$$i(M, N) = \dim_K(\mathrm{Hom}_A(M, N)).$$

Next, we consider the behaviour of intertwining numbers under field extension; this will generalise some of the ideas behind Corollary 8. Our aim will be to refine Corollary 10 to show that, given a group G, some finite extension of $\mathbb{Q}$ is a splitting field, but we shall do more than is required at this stage since we shall return to the question of splitting fields in Chapter 5.

We shall leave to the exercises the computation of intertwining numbers in some special cases. Intertwining numbers will also have a role to play when we consider induced representations in Section 7 and again, though not quite in the form described above, as a central tool in character theory.

Theorem 12. *Let A be a K-algebra and let L be an extension of K. If M and N are A-modules, then*

$$i(M^L, N^L) = i(M, N).$$

Proof. Suppose that M and N afford matrix representations ρ and σ of A respectively. Then a matrix X intertwines ρ and σ if and only if

$$\rho(a)X = X\sigma(a)$$

for all $a \in A$. If this is so, then $\rho^L(1 \otimes a)X = X\sigma^L(1 \otimes a)$ for $a \in A$ and it follows that X intertwines ρ^L and σ^L. Hence, using the identification of Proposition 11, there is a natural injection

$$L \otimes_K \operatorname{Hom}_A(M, N) \to \operatorname{Hom}_{A^L}(M^L, N^L)$$

which is an L-homomorphism and, in particular, $i(M, N) \leqslant i(M^L, N^L)$.

Conversely, suppose that X is a matrix which intertwines ρ^L and σ^L. Then

$$\rho^L(a)X = X\sigma^L(a) \tag{4.4}$$

for each $a \in A^L$. If $a_1, \ldots, a_d$ is a K-basis for A, then all equations of the form (4.4) can be derived as L-linear combinations of those for $a \in \{1 \otimes a_1, \ldots, 1 \otimes a_d\}$ alone. In particular, if $X = (x_{ij})$, then we obtain a system of linear equations in the x_{ij}'s equivalent to the system obtained from the d matrix equations

$$\rho(a_i)X = X\sigma(a_i)$$

for $i = 1, \ldots, d$. The dimension of the solution space is independent of the field (as long as it contains the field of coefficients) and hence, as the coefficients of this system lie in K, we deduce from Proposition 11 that $i(M, N) = i(M^L, N^L)$.

The next result is a special case of the Noether–Deuring theorem: we impose an irreducibility condition which is sufficient for our needs and permits a simplified proof.

Theorem 13. *Let ρ and σ be representations of a K-algebra A, having the same degree d. Let L be an extension of K, and assume that ρ is irreducible. Then $\rho \sim_K \sigma$ if and only if $\rho^L \sim_L \sigma^L$.*

Proof. One direction is clear, so suppose that $\rho^L \sim_L \sigma^L$. If ρ and σ give rise to A-modules M and N respectively, then $i(M^L, N^L) \neq 0$. By Theorem 12, $i(M, N) \neq 0$ and $\mathrm{Hom}_A(M, N) \neq 0$. Since M is irreducible and has the same K-dimension as N, it follows that $M \cong N$; hence $\rho \sim_K \sigma$.

Theorem 13 has been proved without appealing to Schur's lemma or the double centraliser lemma, and thus justifies the remark following Exercise 12 of Section 2 directly, without recourse to the main structure theorems. We shall now apply Theorem 13 to refine Corollary 10.

Let G be a finite group and let ρ be a representation of G over a field K. If ρ is equivalent to a matrix representation in which every element is represented by a matrix with coefficients in a subring R of K, then ρ is said to be *realisable* in R. Clearly this definition cannot be extended to the group ring KG, nor can there be analogues for algebras in general.

Since $\mathbb{A}$ is a splitting field for any finite group, an irreducible representation over $\mathbb{C}$ may be realised over $\mathbb{A}$, and the same is true of any representation by virtue of Maschke's theorem. (See Exercise 12 of Section 2.) The following improvement can be deduced from the detailed structure theorem that we shall prove in the next section, but seems worthy of proof directly from the double centraliser lemma.

Theorem 14. *Let G be a finite group and let K be a field whose characteristic does not divide $|G|$. Suppose that K is a splitting field for G. If F is a subfield of K in which every irreducible representation of G can be realised, then F is also a splitting field for G. In particular, there is an algebraic number field*[†] *which is a splitting field for G.*

Proof. Suppose that ρ is an irreducible representation of G over F which is not absolutely irreducible. Then ρ^K is reducible and, by Maschke's theorem, we may write

$$\rho^K = \sum_{i=1}^{n} \rho_i$$

where $n \geqslant 2$ and $\rho_1, \ldots, \rho_n$ are irreducible representations of G over K. By assumption, $\rho_i \sim_K \sigma_i^K$ for some representation σ_i of G over F. So

$$\rho^K \sim_K \sum_{i=1}^{n} \sigma_i^K = \left(\sum_{i=1}^{n} \sigma_i \right)^K .$$

[†] An algebraic number field is a field which is a finite extension of $\mathbb{Q}$.

Hence by Theorem 13,

$$\rho \sim_F \sum_{i=1}^{n} \sigma_i,$$

contrary to the assumption of irreducibility.

We complete this section with a refinement of a rather different type.

Theorem 15. *Let R be a principal ideal domain and let K be its field of fractions. If G is a group and M is a KG-module, then M contains an RG-submodule N such that $M = KN$ and the rank of N, as a free R-module, is* $\dim_K(M)$. *In particular, the representation afforded by M is K-equivalent to the representation of G afforded by N and so can be realised in R.*

Proof. Pick a K-basis for M, and let N be the RG-submodule that it generates. Then N is finitely generated and free as an R-module. By the cyclic decomposition theorem, N has an R-basis. Clearly $M = KN$ and, since K is the field of fractions of R, an R-basis of N is a K-basis of M.

A case of especial interest is the following. Let G be a group and let K be an algebraic number field which is a splitting field for G. Let R be the ring of algebraic integers in K and let $\mathscr{P}$ be a prime ideal in R containing the rational prime p. Form the localisation $R_{\mathscr{P}}$ of R at $\mathscr{P}$: then $R_{\mathscr{P}}$ is a principal ideal domain with a unique maximal ideal $\tilde{\mathscr{P}}$. Then we have the following.

Corollary 16. *In the above situation, every absolutely irreducible representation of G may be realised in $R_{\mathscr{P}}$.*

This result is the starting point for the study of the relationship between ordinary and modular representation theory, for the quotient $R_{\mathscr{P}}/\tilde{\mathscr{P}}$ is a field $\bar{K}$ of characteristic p. Given an irreducible representation of G over K, we can obtain an irreducible representation over $R_{\mathscr{P}}$ and, viewing this as a matrix representation, we can construct a representation over $\bar{K}$ by reducing the matrix coefficients mod $\tilde{\mathscr{P}}$. This representation will depend on the particular way in which the representation over $R_{\mathscr{P}}$ was obtained: two different constructions may yield inequivalent representations over $\bar{K}$, but it *is* the case that the set of irreducible constituents will be invariant. A proof of this statement is beyond the scope of this book, but we mention it since it underpins the entire Brauer approach to modular representation theory: we shall also have occasion to use Corollary 16 in our study of ordinary characters.

Exercises

1. Let A be a K-algebra and suppose that K is a splitting field for A. Let M and N be irreducible A-modules. Show that
$$\mathrm{Hom}_A(M,N) = \begin{cases} K & \text{if } M \cong N, \\ 0 & \text{otherwise.} \end{cases}$$
2. Let G be a group and let ρ be an absolutely irreducible representation of G. Show that, if ρ is faithful, the centre $Z(G)$ of G is cyclic.
[Note. Use the fact that any finite subgroup of the multiplicative group of a field is cyclic. Also, the conclusion still holds if ρ is just irreducible and faithful: see Exercise 3 of Section 6.]
3. Let A be a simple K-algebra, and let M be a minimal right ideal of A. Show that, if $a \in A$, then either $aM = 0$ or aM is a minimal right ideal of A isomorphic to M as an A-module. Deduce that A is completely reducible as a module over itself.
4. Let A be a K-algebra and suppose that K is a splitting field for A. Let $M_1, \ldots, M_n$ be a complete set of nonisomorphic irreducible A-modules.
   (i) Suppose that M, N are completely reducible A-modules and
   $$M = \oplus r_i M_i \quad \text{and} \quad N = \oplus s_i M_i$$
   where r_i, s_i are nonnegative integers denoting multiplicities. Show that $i(M,N) = \sum r_i s_i$. (In particular, $i(M_i, M) = i(M, M_i) = r_i$.)
   (ii) If M is an arbitrary A-module, prove that
   $$i(M_i, M) = i(M_i, \mathrm{soc}(M)) \quad \text{and} \quad i(M, M_i) = i(M_i, M/\mathrm{rad}(M)).$$
   [See Exercise 3 of Section 3 for the definition of soc (M) and rad (M).]
5. Show that if M and N are completely reducible A-modules, then $i(M,N) = i(N,M)$.
6. Let M, N_1 and N_2 be A-modules. Prove that
$$i(M, N_1 \oplus N_2) = i(M, N_1) + i(M, N_2).$$
7. Let $A = KS_3$. Determine $i(A,A)$ for $K = \mathbb{Z}_2$ and $K = \mathbb{Z}_3$.
[Use computations on the structure of A from Exercise 16 of Section 2.]
8. Deduce from Corollary 8 and Theorem 9 that any extension of a splitting field for a K-algebra is also a splitting field.
[Compare with Exercise 12 of Section 2.]
9. Let R be an arbitrary ring with identity and let M be an irreducible R-module. Show that the endomorphism ring of M is a division ring.
   Show also that, if R is completely reducible as an R-module, then

$R/\mathrm{Ann}(M)$ is a simple ring. Formulate and prove an analogue of Theorem 7 in this case.

5 Semisimple algebras

In this section, we shall obtain results which will enable us to determine the precise structure of a group algebra when the field is a splitting field of characteristic prime to the group order. This will be sufficient for all our subsequent work, although we shall indicate in the exercises the extent to which the results may be directly generalised.

Let A be a K-algebra. Then the radical $J(A)$ annihilates every irreducible A-module and, conversely, every irreducible A-module is isomorphic to a quotient of $A/J(A)$, viewed as an A-module. (See Exercise 10 of Section 2.) Thus, to determine the irreducible A-modules, it is sufficient to consider $A/J(A)$ and we may thus assume that $J(A)=0$. An algebra for which this is so is called *semisimple.* We note that a simple algebra is necessarily semisimple, as is any direct sum of simple algebras. If G is a group whose order is coprime to the characteristic of a field K, then the group algebra KG is completely reducible as a KG-module by Maschke's theorem. Since maximal submodules are maximal right ideals, it follows that $J(KG)=0$ so that KG is semisimple as an algebra.

Let A be a semisimple K-algebra. Then A is completely reducible as an A-module. (See Exercise 3 of Section 3.) Let $\{M_1,\ldots,M_r\}$ be a complete set of nonisomorphic irreducible A-modules and, for $i=1,\ldots,r$, define submodules A_i of A by

$$A_i = \langle M \subseteq A \mid M \cong M_i \rangle. \tag{5.1}$$

Then A is the direct sum of these submodules since they certainly generate A yet have no composition factors in common. The results of Section 4 now permit a precise description of A.

Theorem 17. *Let A be a semisimple K-algebra and let $M_1,\ldots,M_r$ be a complete set of nonisomorphic irreducible A-modules. Define $A_1,\ldots,A_r$ by* (5.1). *Then*

$$A = A_1 \oplus \cdots \oplus A_r \tag{5.2}$$

and the following hold for each i.

- (i) *A_i is a two-sided ideal of A.*
- (ii) *A_i is a simple algebra: in particular, the decomposition* (5.2) *holds as a direct sum of algebras.*
- (iii) *if $D_i = \mathrm{Hom}_K(M_i, M_i)$, then $A_i \cong \mathrm{Hom}_{D_i}(M_i, M_i)$. In particular, if K is a splitting field for A and $\dim_K(M_i) = n_i$, then $A_i \cong \mathcal{M}_{n_i}(K)$.*
- (iv) $\mathrm{Ann}(M_i) = \bigoplus_{j\neq i} A_j$.

Proof. Let M be an irreducible submodule of A_i and let $a \in A$. Then the map $m \to am$ defines an A-module homomorphism from M onto aM. Since M is irreducible, it follows that $aM \subseteq A_i$: hence A_i is a two-sided ideal of A. In particular,

$$A_i A_j \subseteq A_i \cap A_j = 0 \quad \text{if } i \neq j. \tag{5.3}$$

If we now write

$$1 = \sum_i e_i \tag{5.4}$$

where $e_i \in A_i$, then e_i must be an identity for A_i. Thus A_i is a subalgebra of A.

We claim that A_i acts faithfully on M_i. For, if B_i is the kernel of the action, then B_i annihilates every irreducible A-module by virtue of (5.3), and hence $B_i \subseteq J(A) = 0$. (See Exercise 10 of Section 2.) Then, by Lemma 6, A_i is simple, so (i) and (ii) hold. Now Theorem 7, Corollary 8 and Theorem 9 together imply (iii), while (iv) follows immediately from (5.2) and (ii).

The various parts of Theorem 17 are collectively known as the *Wedderburn Theorems*. These hold in slightly greater generality than as stated above, though our approach via the double centraliser lemma has the advantage that it provides exactly the statements that we shall need to develop character theory. However, we shall now see that the usual 'intermediate' results can be obtained as corollaries.

Corollary 18. *Let I be a minimal two-sided ideal of A. Then $I = A_i$ for some i.*

Proof. I is completely reducible as an A-module. For $j = 1, \ldots, r$, put

$$I_j = \langle M \subseteq I \mid M \cong M_j \rangle.$$

Then $I = I_1 \oplus \cdots \oplus I_r$ as a direct sum of modules. However, $I_j \subseteq I \cap A_j$: if $I_j \neq 0$, then $I \cap A_j$ is a nonzero two-sided ideal. Thus $I = I_j$ for some j, and $I = A_j$ by the simplicity of A_j and minimality of I.

Note that this implies that the decomposition (5.2) is unique as a sum of minimal two-sided ideals and hence also as simple subalgebras. In turn, this gives us information about the decomposition (5.4) of the identity.

An element e of A is called an *idempotent* if $e^2 = e \neq 0$. If e lies in the centre $Z(A)$ of A, then e is a *central* idempotent. Two idempotents e, f are *orthogonal* if $ef = fe = 0$, and a central idempotent is *primitive* if it cannot be written as the sum of two orthogonal central idempotents. With these definitions, we can describe the decomposition of 1 as a sum of central primitive idempotents. We shall use these idempotents to derive the fundamental orthogonality relations for complex characters in Chapter 2.

Corollary 19. *Let A be a semisimple K-algebra with a decomposition $A = A_1 \oplus \cdots \oplus A_r$ as a direct sum of simple subalgebras, and let $1 = e_1 + \cdots + e_r$ be the corresponding decomposition of the identity. Then the elements $e_1, \ldots, e_r$ are mutually orthogonal and form precisely the set of central primitive idempotents in A.*

Proof. Since $A_iA_j = 0$ if $i \neq j$, certainly $e_ie_j = 0$, while $e_i = e_i^2$ since e_i is the identity of A_i.

Let f be a central idempotent in A and write

$$f = f_1 + \cdots + f_r$$

where $f_i \in A_i$. Then (5.3) also forces $f_1, \ldots, f_r$ to be pairwise orthogonal central idempotents. If f is primitive, we may suppose that $f = f_1$. Now $A_1 = f_1A$: hence there exists $a \in A_1$ with $e_1 = f_1a$. Then $f_1 = f_1e_1 = f_1(f_1a) = f_1a = e_1$.

Suppose now that K is a splitting field for the semisimple K-algebra A. Then

$$A \cong \bigoplus_{i=1}^{r} \mathcal{M}_{n_i}(K)$$

where $n_1, \ldots, n_r$ are the degrees of the distinct irreducible representations of A. Thus $n_1, \ldots, n_r$ are uniquely determined, though they are, of course, invariants by consideration of dimension. The centre of $\mathcal{M}_{n_i}(K)$ consists of the scalar matrices. Thus we have the following.

Corollary 20. *If K is a splitting field for the semisimple K-algebra A, then*

(i) $\dim_K(Z(A)) = r$, *the number of inequivalent absolutely irreducible representations of A,*
(ii) *the right regular representation $\sim \sum n_i\rho_i$, and*
(iii) $\sum_{i=1}^{r} n_i^2 = \dim_K(A)$.

In the case of group algebras, we can say more.

Theorem 21. *Let G be a finite group and let K be an arbitrary field. Then the set of sums of elements in each conjugacy class of G forms a basis for $Z(KG)$. If K is a splitting field for G of characteristic not dividing the order of G, then the number of distinct irreducible representations of G over K is equal to the number of conjugacy classes of elements of G.*

Proof. Suppose that $\sum a_g g \in Z(KG)$. If $x \in G$, then

$$x(\sum a_g g)x^{-1} = \sum a_g(xgx^{-1}) = \sum a_g g:$$

hence $a_g = a_{x^{-1}gx}$ for all $x, g \in G$. Thus the set of conjugacy class sums spans $Z(KG)$. On the other hand, these sums clearly form a linearly independent set. The second statement follows from Corollary 20.

This theorem, together with some additional information that can be derived from Theorem 17 will form the starting point in Chapter 2 for our development of character theory.

Exercises

1. Let D be a division ring and let V be a finite dimensional right vector space over D. Let D° be the opposite ring to D (that is, with a multiplication defined by $a \circ b = ba$). Show that $\mathrm{Hom}_D(V, V) \cong \mathscr{M}_n(D^\circ)$ as a ring, where $n = \dim_D(V)$. Deduce that, if A is a simple K-algebra and M is an irreducible A-module with $\mathrm{Hom}_A(M, M) = D$, then $A \cong \mathscr{M}_n(D^\circ)$ where n is the number of composition factors of A as an A-module.
   [Note. The uniqueness of n and D in the isomorphism $A \cong \mathscr{M}_n(D^\circ)$ will be established in Exercise 4.]
2. Let D be a division ring and let $R = \mathscr{M}_n(D)$. Let M be a minimal right ideal of R. Show that there exists an element $r \in R$ such that rM consists of those matrices whose entries outside the first row are all zero. Deduce that R is a simple ring and that R is completely reducible as an R-module. Hence show that all irreducible R-modules are isomorphic.
3. Let R be a simple ring with identity. Let e be an idempotent in R and put $M = eR$. Prove that, if $\alpha \in \mathrm{Hom}_R(M, M)$, then
   $$x\alpha = (e\alpha)x = (e(e\alpha)e)x \quad \text{for all } x \in M.$$
   Hence prove that $\mathrm{Hom}_R(M, M) \cong (eRe)^\circ$ where $^\circ$ denotes the opposite ring, as in Exercise 1.
4. Let R and M be as in Exercise 2. Show, by using the result of Exercise 3, that $\mathrm{Hom}_R(M, M) \cong D^\circ$. Hence prove that, if D and D' are division rings for which $\mathscr{M}_n(D) \cong \mathscr{M}_{n'}(D')$ for some n, n', then $n = n'$ and $D \cong D'$.
   [Hint. Find a suitable idempotent in rM to characterise D°; then count composition factors to show that $n = n'$ and characterise $(D')^\circ$ via the isomorphism $R \cong \mathscr{M}_n(D')$.]
5. Let R be a ring with identity which satisfies the descending chain condition on right ideals. Assume (or prove) that R then satisfies the ascending chain condition also. Using this as a finiteness condition in place of dimension, obtain an analogue of Theorem 17: namely, prove that a semisimple ring with identity satisfying the descending

chain condition on right ideals can be expressed uniquely as the direct sum of matrix rings over division rings.
[This is a stronger version of the Wedderburn theorems.]

6. Let A be a K-algebra and let $\rho_1, \ldots, \rho_r$ be its distinct inequivalent irreducible representations. For $i = 1, \ldots, r$, define the trace function $t_i: A \to K$ by $t_i(a) = \mathrm{tr}(\rho_i(a))$. Prove that, if A is semisimple, then the functions $t_1, \ldots, t_r$ are linearly independent over K, and deduce that the same is true in general.
7. Prove that the number of inequivalent irreducible representations of a group over an arbitrary field cannot exceed the number of conjugacy classes. Can this bound be improved if the field has characteristic dividing the group order?
[Hint. Recall that the functions t_i of Exercise 6 are constant on conjugacy classes.]
8. Let A be a K-algebra and suppose that K is a splitting field for A. Let L be an extension of K. Prove that $J(A^L) \supseteq L \otimes_K J(A)$ and deduce that, if ρ and σ are irreducible representations of A (over K), then $\rho \sim_K \sigma$ if and only if $\rho^L \sim_L \sigma^L$. Show also that the inclusion may be strengthened to equality.
[Compare with Theorem 13.]
9. Let A be a semisimple K-algebra and let $\{M_1, \ldots, M_r\}$ be a complete set of nonisomorphic irreducible A-modules. Let $D_i = \mathrm{Hom}_A(M_i, M_i)$. Prove that D_i is finite dimensional as an algebra over K and that
$$\dim_K(A) = \sum_{i=1}^{r} \dim_K(D_i) \cdot (\dim_{D_i}(M_i))^2.$$
10. Let $\mathbb{H}$ be the algebra of Hamilton's quaternions, namely the four-dimensional real algebra with basis $1, i, j, k$ and multiplication defined by
$$i^2 = j^2 = k^2 = -1, \quad ij = -ji = k, \quad jk = -kj = i, \quad ki = -ik = j.$$
Show that $\mathbb{H}^{\mathbb{C}}$ is noncommutative, and hence that $\mathbb{H}$ has a two-dimensional representation over $\mathbb{C}$. Find an explicit representation.
Show also that $\mathbb{H}$ is a *division algebra* over $\mathbb{R}$, and deduce that it has a unique representation (of degree 4) over $\mathbb{R}$.
11. Let Q_8 be the quaternion group of order 8. Show that there is an $\mathbb{R}$-algebra homomorphism from $\mathbb{R}Q_8$ onto $\mathbb{H}$. Deduce that Q_8 has five inequivalent irreducible representations over $\mathbb{R}$, four of degree 1 and one of degree 4.

6 Clifford's theorem

For the remainder of this chapter, we will restrict our attention to the representations of groups and study, in particular, the relationships between the representations of a group and those of its subgroups. These results will be independent of the structure theorems proved in the previous two sections.

Let K be an arbitrary field and let G be a group and H a subgroup of G. In general, little can be said about the structure of a KG-module when viewed as a KH-module, but an important special case occurs when H is a normal subgroup. We shall consider this now in view of its important applications to the internal study of finite groups: as an example of such an application, we shall give a proof of the Frobenius Conjecture that a finite group admitting a fixed-point-free automorphism of prime order is nilpotent.

Suppose that $H \trianglelefteq G$ and that M is a KH-module. Regarding M as a vector space only, we can define a new action of H in the following way. Fix $g \in G$ and, for $m \in M$ and $h \in H$, let

$$m \cdot h = m(ghg^{-1})$$

where the action on the right-hand side is that of the original module M. Then it is easily verified that this gives M the structure of a KH-module when extended by linearity (in general different from the original); this is called a *conjugate* of M and is denoted by $M^{(g)}$. The same term will be applied to the corresponding representations also and, in view of the following lemma, conjugacy need be considered only up to equivalence.

Lemma 22. *If M and N are isomorphic KH-modules, then $M^{(g)}$ and $N^{(g)}$ are KH-isomorphic for all $g \in G$. The map $M \to M^{(g)}$ induces a transitive permutation representation of G on the set of conjugates of a fixed KH-module.*

Proof. Let $\varphi : M \to N$ be a KH-isomorphism. Since

$$(m \cdot h)\varphi = (mghg^{-1})\varphi = (m\varphi)ghg^{-1} = (m\varphi) \cdot h$$

for all $m \in M$ and $h \in H$, φ is also a KH-isomorphism from $M^{(g)}$ onto $N^{(g)}$. For the second statement, we need only show[†] that, if $g_1, g_2 \in G$, then

$$M^{(g_1 g_2)} \cong_{KH} (M^{(g_1)})^{(g_2)},$$

and this is clear since

$$m((g_1 g_2)h(g_1 g_2)^{-1}) = m(g_1(g_2 h g_2^{-1})g_1^{-1})$$

for $m \in M$ and $h \in H$.

[†] By the symbol $\cong_{KH}$, we shall mean isomorphism as KH-modules. In general, we shall specify the type of isomorphism in this way only in cases of ambiguity–for example, $\cong_K$ will denote isomorphism as vector spaces.

As a consequence of this lemma, we may define the *inertia group* $I(M)$ of a KH-module M as the stabiliser of M under this permutation action. In particular, $M^{(g)}$ will depend only on the coset $I(M)g$. We now come to the main result of this section.

Theorem 23 (Clifford, 1937). *Let M be an irreducible KG-module and let H be a normal subgroup of G. Then M_H is completely reducible as a KH-module and its irreducible KH-submodules are conjugate (up to isomorphism).*

Proof. Let L be a KH-submodule of M_H. If $g \in G$, then Lg is also a KH-submodule and, since $L = (Lg)g^{-1}$, we have $\dim_K(L) = \dim_K(Lg)$. If L is irreducible, then so is Lg, and

$$M_H = \sum_{g \in G} Lg$$

since $\sum Lg$ is a G-invariant subspace of M. Hence M_H is completely reducible by Proposition 3. For $g \in G$, define a map $\varphi: L^{(g)} \to Lg$ by putting $l\varphi = lg$ for all $l \in L$. Since

$$(l \cdot h)\varphi = (lghg^{-1})\varphi = (lghg^{-1})g = (lg)h$$

for all $l \in L$ and $h \in H$, it follows that φ is a KH-homomorphism. Now $\ker \varphi = 0$ so that $Lg \cong L^{(g)}$, while any composition factor of M_H is necessarily isomorphic to some Lg.

Since M_H has only finitely many composition factors, for each irreducible submodule we may consider the sum of all submodules isomorphic to it. The distinct such sums, $M_1, \ldots, M_t$, are called the *homogeneous components* of M_H and, since M_H is completely reducible, $M_H = M_1 \oplus \cdots \oplus M_t$. In view of Lemma 22, Clifford's theorem has an immediate refinement.

Corollary 24. *With the notation above, G acts as a transitive permutation group on the homogeneous components under the action $g: M_i \to M_i g$ for all $g \in G$. Furthermore, $M_H \cong e(L^{(g_1)} \oplus \cdots \oplus L^{(g_t)})$ where $\{L^{(g_1)}, \ldots, L^{(g_t)}\}$ is a complete set of nonisomorphic conjugates of a fixed irreducible submodule L of M_H, $t = [G:I(L)]$ and e is a nonnegative integer.*

Proof. Let L be an irreducible KH-submodule of M_H. Then

$$M_H = \sum_{g \in G} Lg.$$

Since $Lg \cong_{KH} L^{(g)}$, the first statement follows from Lemma 22. The second is simply the assertion that each homogeneous component has the same number of composition factors.

The number e in Corollary 24 is called the *ramification index* of M with respect to H. In general, it is very difficult to determine, and we shall return to a further discussion of Clifford's theorem at the end of the next section.

As an application of Clifford's theorem (and, indeed, Maschke's theorem), we shall now give a proof of the Frobenius Conjecture. The proof falls into two parts. The first is to show that solubility implies nilpotence; this is usually ascribed to Witt, and the proof shows the close relationship between pure group theory and representation theory as indicated in the first two exercises of Section 2 and which culminated in the famous Hall–Higman theorem (1956). The second, by far the deeper, is to establish solubility. This was first proved by Thompson (1959) using his normal p-complement theorem (1960) which itself was a massive argument based on some of the ideas of the Hall–Higman paper. Thompson (1964) subsequently improved his normal p-complement theorem and we will state it as Theorem 25 in this form. We shall not give a proof here since it lies outside the spirit of this book: much consists of the systematic reduction of a minimal counterexample, starting with Frobenius' normal p-complement theorem. There are also a number of variants. (See, for example, Gorenstein [G; p. 280].)

For a p-group P, let $m(P)$ denote the maximum of the ranks of the abelian subgroups of P, and define the *Thompson subgroup*

$$J(P) = \langle A \mid A \text{ is an abelian subgroup of } P \text{ of rank } m(P) \rangle.$$

Notice, in particular, that $J(P)$ is a characteristic subgroup of P and that, if $J(P) \subseteq P^* \subseteq P$, then $J(P^*) = J(P)$.

Theorem 25. *Let G be a finite group. Suppose that p is an odd prime divisor of $|G|$ and that P is a Sylow p-subgroup of G. If $N_G(J(P))$ and $C_G(Z(P))$ have normal p-complements, then G has a normal p-complement.*[†]

Theorem 26. *Let G be a finite group which admits a fixed-point-free automorphism α of prime order p. Then G is nilpotent.*

Proof. First, assume that G is soluble and that G is of minimal order

[†] A group G is said to have a normal p-complement if it has a normal p'-subgroup N such that $G = PN$ for a Sylow p-subgroup P of G.

subject to not being nilpotent. Note that $|G| \equiv 1 \pmod p$. If M and N are α-invariant subgroups of G with $N \lhd M$, then α cannot fix a coset of N in M without fixing an element in it. So every proper α-invariant section of G admits α fixed-point-freely, and hence is nilpotent.

Let $F(G)$ be the Fitting subgroup of G; that is, the maximal nilpotent normal subgroup of G. Let $\bar{G} = G/F(G)$. Then the minimality of G forces $\bar{G}$ to be an elementary abelian q-group, for some prime q, on which α acts irreducibly. Now, if Q_0 were a nonidentity Sylow q-subgroup of $F(G)$, the quotient G/Q_0 would be nilpotent, as then would be G. So q does not divide $|F(G)|$. Let Q be an α-invariant Sylow q-subgroup of G; such exists since $p \nmid |G|$. For some prime r, there is a Sylow r-subgroup R of $F(G)$ such that $[Q, R] \neq 1$; then $G = QR$ by minimality. By the Hall–Burnside theorem, Q must act nontrivially on the Frattini quotient $R/\Phi(R)$; hence R must be elementary abelian by the minimality of G. Let H denote the natural semidirect product $\langle \alpha \rangle Q$; then Q is the unique minimal normal subgroup of H, and H acts faithfully on R.

We now view R as a $\mathbb{Z}_r H$-module and apply Clifford's theorem; by the minimality of $|G|$ and Maschke's theorem, we may suppose that the action is faithful and irreducible. Since α acts fixed-point-freely on R, no eigenvalue of α can be 1, and the same is true if the field is extended. Adjoin a primitive qth root ω of unity to $\mathbb{Z}_r$ and put $K = \mathbb{Z}_r(\omega)$. Then K is a splitting field for Q. (Exercise 14 of Section 2). Let M be an irreducible KH-submodule of R^K. By Corollary 24, M_Q has either one or p homogeneous components for Q. If the number were p, we could put $M_Q = M_1 \oplus \cdots \oplus M_p$, choose $m \in M_1 - \{0\}$, and observe that α fixed the element $m + m\alpha + \cdots + m\alpha^{p-1}$; so there can be only one homogeneous component. Now M_Q is a direct sum of isomorphic, absolutely irreducible KQ-modules. Thus Q acts on M as scalar multiplication and the action commutes with that of α. Since R^K is completely reducible as a KH-module, the actions of Q and $\langle \alpha \rangle$ on R^K commute and hence they commute on R also. But we may view H as a subgroup of the semidirect product $G\langle \alpha \rangle$ so that $[\alpha, Q] \subseteq R \cap Q = 1$. This is false. Hence, if G is soluble, then G is nilpotent.

To complete the proof of Theorem 26 using Thompson's normal p-complement theorem, let G be a counterexample of minimal order to solubility. Then G can have no proper, nonidentity, normal α-invariant subgroups. G cannot be of prime power order. Let q be an odd prime divisor of $|G|$ and let Q be an α-invariant Sylow q-subgroup of G. Then $N_G(J(Q))$ and $C_G(Z(Q))$ are proper α-invariant subgroups of G and so are nilpotent. By Thompson's theorem, G has a normal q-complement which is α-invariant, contrary to the above. This completes the proof.

Exercises

1. Show that every irreducible representation of a p-group over a field of characteristic p is trivial.
[Hint. Assume faithful and restrict to a cyclic subgroup in the centre.]
2. Show that any normal p-subgroup of a group lies in the kernel of every irreducible representation over a field of characteristic p.
3. Suppose that G has a faithful irreducible representation over some field. Show that $Z(G)$ is cyclic.
4. Let P be a p-group and let Q be a subgroup of index p in P. Suppose that P contains an element α of order p such that $P = Q\langle\alpha\rangle$ and $|C_Q(\alpha)| = p$. Show that, if P has a faithful representation in which α acts fixed-point-freely, then $|Q| = p$.

7 Induced representations

Given a group G and a subgroup H, we shall associate with each representation of H a representation of G, called the *induced representation*. This will provide an important connection between the representations of a finite group and the representations of its subgroups. After developing basic properties, we shall study some intertwining numbers and derive relations due to Nakayama. In the case of completely reducible modules, these simplify so that when we study the character theory of finite groups over $\mathbb{C}$, we can see how induction plays the role of the adjoint of restriction. In fact, we shall use that idea as the definition of an induced character and show in the next chapter that the result coincides with what would have been obtained starting from an induced representation. In this section, we shall develop the theory of induced representations over arbitrary fields. After obtaining the Nakayama relations, we shall prove an important theorem due to Mackey (1951) and apply it to prove a result about permutation representations and also to study induction from a normal subgroup; in the latter application, we shall see a similarity with some of the ideas of the previous section and we will apply Clifford's theorem to show that, under suitable hypotheses, an irreducible representation induces an irreducible representation. This final result will be of central importance in the later work of this book.

Although the initial definition (7.1) extends naturally to an algebra and a subalgebra, this seems to have no useful application and the subsequent development of the theory depends on starting with a group and its subgroups. So, throughout this section, let G be a group, H a subgroup of G, and K an arbitrary field. Let $\{g_1, \ldots, g_n\}$ be a (right) transversal of H in G; then $G = \bigcup Hg_i$. The group algebra KG has the structure of a free *left* KH-module since

$$KG = \bigoplus_{i=1}^{n} KHg_i.$$

Let M be a right KH-module affording a representation ρ of H. Define a right KG-module M^G by forming the tensor product

$$M^G = M \otimes_{KH} KG. \tag{7.1}$$

This definition is strictly formal, but as an alternative one may construct M^G by using the properties of Lemma 27 below and verifying that one has the structure of a right KG-module. The module M^G is said to have been *induced by* M, or induced *from* the subgroup H, and the representation ρ^G that M^G affords is called the *induced representation.* The construction given above does not depend on the particular transversal taken: the structure with regard to a fixed transversal is described in the following manner and we shall go on to discuss the similarity of matrix representations afforded by different transversals.

Lemma 27. *With the above notation, let*

$$M_i = M \otimes g_i = \{m \otimes g_i | m \in M\}, \quad i = 1, \ldots, n.$$

Then $M^G = M_1 \oplus \cdots \oplus M_n$ *as a vector space over* K. *If* $g \in G$ *and* $g_i g = hg_j$, *then* $M_i g \subseteq M_j$ *and* $(m \otimes g_i)g = mh \otimes g_j$.

Proof. The first statement is obvious. Now

$$(m \otimes g_i)g = m \otimes (g_i g) = m \otimes (hg_j) = mh \otimes g_j,$$

and the inclusion follows.

Since an element g acts as a nonsingular linear transformation on M^G and the subspaces M_i have the same dimension, it follows that the inclusion of the lemma may be replaced by equality. Thus the action of an element g on M^G may be viewed as first inducing the same permutation of the subspaces M_i as on the right cosets of H, and then an action by elements of H on each. This gives rise to the following precise matrix representation.

Theorem 28. *With the hypothesis and notation above, let* $m_1, \ldots, m_t$ *be a* K*-basis for* M. *Let* ρ *be the matrix representation of* H *afforded by* M *with respect to this basis. Then, with respect to the ordered basis*

$$m_1 \otimes g_1, \ldots, m_t \otimes g_1, m_1 \otimes g_2, \ldots, m_t \otimes g_n,$$

$\rho^G(g)$ *is a matrix having block form* $(R_{ij}(g))$ $(1 \leqq i, j \leqq n)$ *where*

$$R_{ij}(g) = \begin{cases} \rho(g_i g g_j^{-1}) & \textit{if } g_i g g_j^{-1} \in H, \\ 0 & \textit{otherwise.} \end{cases}$$

Each row and column of blocks contains exactly one nonzero block, and each nonzero block is a nonsingular $t \times t$ matrix.

Proof. From the remark following Lemma 27, it is clear that $\rho^G(g)$ has such a block form and that $R_{ij}(g) \neq 0$ if and only if $M_i g \subseteq M_j$. Now, if $g_i g = h g_j$ and $m \in M$,

$$(m \otimes g_i)g = m(g_i g g_j^{-1}) \otimes g_j,$$

and the result follows.

It is immediately clear from the discussion that, in the special case that M is the trivial module, we obtain a permutation module.

Corollary 29. *If 1_H denotes the trivial module for KH, then $(1_H)^G$ affords the permutation representation of G on the right cosets of H.*

A further consequence of the matrix formulation of Theorem 28 is that the coefficients of the matrix $\rho^G(g)$ lie in the field generated by the coefficients of the matrices $\rho(h)$ for $h \in H$. Thus, viewing realisation as a process involving a change of basis in the underlying vector space, we can deduce the following corollary.

Corollary 30. *Let G be a group and let H be a subgroup of G. If K is an arbitrary field and ρ is a representation of H over K which can be realised over a subfield F, then ρ^G can be realised over F also.*

We note that it is easy to verify directly that the conclusion of Theorem 28 does define a matrix representation of G. Also, if $\{h_1 g_1, \ldots, h_n g_n\}$ is another transversal giving rise to a matrix representation ρ', and $g_i g g_j^{-1} \in H$, then

$$\rho((h_i g_i) g (h_j g_j)^{-1}) = \rho(h_i) \rho(g_i g g_j^{-1}) \rho(h_j)^{-1},$$

so that ρ and ρ' are similar under conjugation by the matrix with blocks of the form $\rho(h_i)$ on the diagonal, while permuting the members of a transversal will correspond simply to conjugation by the corresponding permutation matrix.

We next examine some of the basic properties of induction. The first result follows from a formal calculation using (7.1), and the proof is left as an exercise.

Proposition 31. *Let H_1 and H_2 be subgroups of a group G with $H_1 \subseteq H_2$, and let M and M' be KH_1-modules. Then*

(i) $(M \oplus M')^G \cong M^G \oplus M'^G$, *and*
(ii) $(M^{H_2})^G \cong M^G$.

In the proofs of the next two propositions, we take the same notation as in Lemma 27.

Proposition 32. *Let M be a KH-module and let N be a KG-module. Then $(M^*)^G \cong (M^G)^*$ and $N \otimes M^G \cong (N_H \otimes M)^G$.*

Proof. The contragredient module M^*, as defined in Section 2, has its structure given by

$$m(m^*g) = (mg^{-1})m^*$$

for $m \in M$, $m^* \in M^*$ and $g \in G$. Define a map $\varphi: (M^*)^G \to (M^G)^*$ by letting

$$(m \otimes g_k)[(m^* \otimes g_i)\varphi] = (\delta_{ik} m)m^*$$

and extending by linearity. By taking a K-basis for M and its dual basis in M^*, it can be seen that φ is nonsingular: φ is then a K-isomorphism since the dimensions are the same. It follows that φ is a KG-isomorphism since, if $g_i g = hg_j$,

$$\begin{aligned}(m \otimes g_k)[((m^* \otimes g_i)g)\varphi] &= (m \otimes g_k)[(m^*h \otimes g_j)\varphi] \\ &= \delta_{jk} m(m^*h) \\ &= \delta_{jk}(mh^{-1})m^*\end{aligned}$$

while

$$\begin{aligned}(m \otimes g_k)[(m^* \otimes g_i)\varphi \cdot g] &= [(m \otimes g_k)g^{-1}](m^* \otimes g_i)\varphi \\ &= \begin{cases} 0 & \text{if } k \neq j, \\ (mh^{-1})m^* & \text{if } k = j. \end{cases}\end{aligned}$$

For the second isomorphism, we proceed similarly. We may obtain a K-isomorphism $\psi: N \otimes M^G \to (N_H \otimes M)^G$ by defining

$$[n \otimes (m \otimes g_i)]\psi = (ng_i^{-1} \otimes m) \otimes g_i$$

for $n \in N$ and $m \in M$, and extending linearly. Then ψ is also a KG-isomorphism since, if $g \in G$ and $g_i g = hg_j$,

$$\begin{aligned}([n \otimes (m \otimes g_i)]\psi)g &= ((ng_i^{-1} \otimes m) \otimes g_i)g \\ &= [(ng_i^{-1} \otimes m)h] \otimes g_j \\ &= [ng_i^{-1}h \otimes mh] \otimes g_j\end{aligned}$$

while

$$\begin{aligned}[(n \otimes (m \otimes g_i))g]\psi &= (ng \otimes (mh \otimes g_j))\psi \\ &= (ngg_j^{-1} \otimes mh) \otimes g_j.\end{aligned}$$

In the next proposition, we compute some intertwining numbers using the notation i_G and i_H for the intertwining numbers of KG- and KH-modules respectively. We recall that K may be viewed as the trivial module.

Proposition 33. *Let H be a subgroup of G and let M be a KH-module. Then $i_H(K, M) = i_G(K, M^G)$.*

Proof. With the notation of Lemma 27, G permutes the subspaces $\{M_i\}$ transitively, so that a vector $\hat{m} \in M^G$ is left invariant by G if and only if it is of the form

$$\hat{m} = \sum_{i=1}^{n} m \otimes g_i$$

where $mh = m$ for all $h \in H$. The result follows.

Lemma 34. *Let G be a group and let L and M be KG-modules. Let N be the submodule of fixed points of G in $L^* \otimes_K M$. Then*

$$i(L, M) = \dim_K(N) = i(K, L^* \otimes_K M).$$

Proof. For $\psi \in L^*$ and $m \in M$, define a map $\psi \circ m: L \to M$ by the formula $l(\psi \circ m) = (l\psi)m$. Then $\psi \circ m \in \mathrm{Hom}_K(L, M)$ and, by taking a basis for L, the dual basis for L^* and a basis for M, it can be easily checked that the map

$$\psi \otimes m \to \psi \circ m$$

extends to a well-defined K-isomorphism from $L^* \otimes_K M$ onto $\mathrm{Hom}_K(L, M)$.

We must show that N corresponds under this isomorphism to $\mathrm{Hom}_{KG}(L, M)$. Now $\sum \psi_i \circ m_i \in \mathrm{Hom}_{KG}(L, M)$ if and only if

$$[l(\sum \psi_i \circ m_i)]g = lg(\sum \psi_i \circ m_i)$$

for all $l \in L$ and $g \in G$. But this is the case if and only if

$$\begin{aligned} l(\sum \psi_i \circ m_i) &= [lg(\sum \psi_i \circ m_i)]g^{-1} \\ &= \sum_i [((lg)\psi_i)m_i]g^{-1} \\ &= \sum_i (l(\psi_i g^{-1}))(m_i g^{-1}) \\ &= l\left(\sum_i (\psi_i g^{-1}) \circ (m_i g^{-1})\right): \end{aligned}$$

that is, if and only if $\sum \psi_i \otimes m_i \in N$.

We are now in a position to derive the Nakayama relations. If the characteristic of K does not divide the group order, the two parts become the same statement since $i(M, N) = i(N, M)$ if M and N are completely reducible modules. In the case that $K = \mathbb{C}$, these relations (or, strictly speaking, their character theoretic equivalent) are known as Frobenius' reciprocity theorem, and it is quite easy to derive the result character theoretically. However, as we commented before, in the next chapter we shall use this as our fundamental definition.

Theorem 35. *Let H be a subgroup of the group G. Let M be a KH-module and let N be a KG-module. Then*

(i) $i_G(N, M^G) = i_H(N_H, M)$, *and*
(ii) $i_G(M^G, N) = i_H(M, N_H)$.

Proof. Noting that $(N^*)_H \cong (N_H)^*$, and applying Propositions 32, 33 and Lemma 34, we see that

$$\begin{aligned} i_G(N, M^G) &= i_G(K, N^* \otimes M^G) \\ &= i_H(K, (N^*)_H \otimes M) \\ &= i_H(K, (N_H)^* \otimes M) \\ &= i_H(N_H, M) \end{aligned}$$

and that

$$\begin{aligned} i_G(M^G, N) &= i_G(K, (M^G)^* \otimes N) \\ &= i_G(K, (M^*)^G \otimes N) \\ &= i_H(K, M^* \otimes N_H) \\ &= i_H(M, N_H). \end{aligned}$$

We now turn to a discussion of Mackey's theorem. We return to the situation at the beginning of the section and, in particular, the notation of Lemma 27. Consider the (vector space) summand $M_i = M \otimes g_i$. In a natural way, this has the structure of a KH^{g_i}-module, where $H^{g_i} = g_i^{-1} H g_i$, via the action

$$(m \otimes g_i)(g_i^{-1} h g_i) = mh \otimes g_i:$$

if $H \trianglelefteq G$, it can be seen that M_i may be identified with the conjugate $M^{(g_i)}$ of M defined in the previous section.

Theorem 36 (Mackey Decomposition Theorem). *Let H_1 and H_2 be subgroups of G and let M be a KH_1-module. Suppose that $\{x_1, \ldots, x_m\}$ is a set of (H_1, H_2)-double coset representatives in G. Then*

$$(M^G)_{H_2} \cong \bigoplus_{i=1}^{m} ((M \otimes x_i)_{H_1^{x_i} \cap H_2})^{H_2}.$$

Proof. Let $\{g_1, \ldots, g_n\}$ be a transversal for H_1 in G. The subgroup H_2 permutes the cosets of H_1 in orbits consisting of the cosets lying in each double coset $H_1 x_i H_2$: thus, as a KH_2-module, there is a direct sum decomposition

$$(M^G)_{H_2} = W_1 \oplus \cdots \oplus W_m$$

where

$$W_i = \bigoplus_{g_j \in H_1 x_i H_2} M_j.$$

For any fixed i, let $\{y_1, \ldots, y_l\}$ be a transversal for $H_1^{x_i} \cap H_2$ in H_2. Then $H_1 x_i y_1, \ldots, H_1 x_i y_l$ is the set of right cosets of H_1 contained in $H_1 x_i H_2$ and

$$W_i = \bigoplus_{k=1}^{l} M \otimes (x_i y_k) = \bigoplus_{k=1}^{l} (M \otimes x_i) y_k.$$

However, $(M \otimes x_i)$ may be viewed as a $KH_1^{x_i}$-module and, in particular, as a $K(H_1^{x_i} \cap H_2)$-module. Thus, by definition,

$$W_i \cong ((M \otimes x_i)_{H_1^{x_i} \cap H_2})^{H_2}.$$

We observe that Mackey's theorem provides a means of computing the action of G on induced modules from a suitable knowledge of intersections of subgroups and their conjugates. This takes place, for example, if G is a doubly transitive permutation group and $H_1 = H_2$ is the stabiliser of a point. More generally, and this has been a major application, one has good control in groups with a BN-pair. The general study of the representations of such groups is beyond the scope of this book (and we refer the reader to the book by Carter [C]), but we shall use Theorem 36 to compute the characters of the groups $\mathrm{SL}(2, 2^n)$ as an example in Chapter 2.

As our first application of Mackey's theorem, we shall use the Nakayama relations to deduce the Intertwining Number Theorem. Let H_1 and H_2 be two subgroups of G and let M_1 and M_2 be KH_1- and KH_2-modules, respectively. We wish to obtain a formula for $i(M_1^G, M_2^G)$ which can be computed from a knowledge of the intersections of the conjugates of H_1 and H_2. For the next theorem, we fix the following notation. $\mathscr{D}$ will be the set of (H_1, H_2)-double cosets in G and, if $x, y \in G$, we put $M_1^{[x]} = (M_1 \otimes x)_{H_1^x \cap H_2^y}$ and $M_2^{[y]} = (M_2 \otimes y)_{H_1^x \cap H_2^y}$.

Theorem 37 (Intertwining Number Theorem). *Assume the situation of the paragraph above. Let $D \in \mathscr{D}$. If $xy^{-1} \in D$, then $i(M_1^{[x]}, M_2^{[y]})$ is independent of the particular choice of x and y. If, for any such x, y, we put $i(M_1, M_2, D) = i(M_1^{[x]}, M_2^{[y]})$, then*

$$i(M_1^G, M_2^G) = \sum_{D \in \mathscr{D}} i(M_1, M_2, D).$$

Proof. By Theorems 35 and 36, we have the formulae

$$\begin{aligned} i_G(M_1^G, M_2^G) &= i_{H_2}((M_1^G)_{H_2}, M_2) \\ &= i_{H_2}\left(\left(\bigoplus_{i=1}^{m} (M_1 \otimes x_i)_{H_1^{x_i} \cap H_2}\right)^{H_2}, M_2\right) \\ &= \sum_i i_{H_2}(((M_1 \otimes x_i)_{H_1^{x_i} \cap H_2})^{H_2}, M_2) \\ &= \sum_i i_{H_1^{x_i} \cap H_2}((M_1 \otimes x_i)_{H_1^{x_i} \cap H_2}, (M_2)_{H_1^{x_i} \cap H_2}). \end{aligned}$$

Letting $D = H_1 x_i H_2$, putting $xy^{-1} = x_i$ and conjugating by y, the ith term of this sum becomes

$$i_{H_1^x \cap H_2^y}(M_1^{[x]}, M_2^{[y]}) = i(M_1, M_2, D),$$

establishing the independence from the choice of x, y also.

As an application, let G be a transitive permutation group and let H be the stabiliser of a point. Then the *rank* of G as a permutation group is the number of orbits of H, including the fixed point; in particular, the rank of a doubly transitive group is 2. Take $H_1 = H_2 = H$ and $M_1 = M_2 = 1_H$. Then it is clear that $i(M_1, M_2, D) = 1$ for every double coset. However, the number of double cosets is just the rank. So we can state the following.

Corollary 38. *Let G be a finite transitive permutation group and let H be the stabiliser of a point. Then $i((1_H)^G, (1_H)^G)$ is equal to the rank of G as a permutation group.*

We complete this section with a discussion of induction from a normal subgroup and with ideas related to Clifford's theorem. Let H be a normal subgroup of G and take $H = H_1 = H_2$ in Mackey's Theorem. Then the double coset decomposition becomes a single coset decomposition

$$G = \bigcup Hg_i,$$

where $\{g_1, \ldots, g_n\}$ is a transversal for H in G. The following result and especially its consequence, Corollary 40, have immediate character theoretic translations and will be used in Theorem 2.43 to determine the characters of Frobenius groups which will have a major role to play in the remainder of this book.

Theorem 39. *Let H be a normal subgroup of G. Let M be a KH-module and let N be the direct sum of the distinct conjugates of M. Then $(M^G)_H$ is isomorphic to a direct sum of r copies of N, where $r = [I(M), H]$ and $I(M)$ is the inertia group of M.*

Proof. Let $\{x_1, \ldots, x_r\}$ be a transversal for H in $I(M)$ and let $\{y_1, \ldots, y_s\}$ be a transversal for $I(M)$ in G. Then $\{x_i y_j\}$ is a transversal for H in G. Now Theorem 36 gives the isomorphism

$$(M^G)_H \cong \bigoplus_{i,j} M^{(x_i y_j)}$$

using the identification of $M \otimes g$ with the conjugate $M^{(g)}$, while $M^{(x_i y_j)} \cong M^{(x_m y_n)}$ if and only if $j = n$, by Lemma 22.

In general, it is difficult to say much directly about the structure of the induced module M^G: if $I(M) = G$, the module M is said to be *G-stable* and in this case the situation is particularly complex. An elaborate axiomatic theory of Clifford extensions was developed by Dade (1970) and recast by Cline (1972) in terms of graded algebras; this was reformulated by Dade (1980) and the interested reader is referred also to Curtis and Reiner [CRI; §11C] for an overall discussion of the construction of various graded algebras. Even in the special case where K is a splitting field, the determination of the structure of M^G involves the study of projective representations. In the opposite extreme case, however, where $I(M) = H$, the structure of M^G may be described more easily, at least if M is irreducible.

Corollary 40. *Assume the hypothesis of Theorem* 39. *Suppose, further, that M is irreducible. Then M^G is irreducible if $I(M) = H$. Conversely, if M^G is absolutely irreducible, then $I(M) = H$.*

Proof. Suppose that $I(M) = H$ and that $(M^G)_H \cong M_1 \oplus \cdots \oplus M_s$ where $M_1, \ldots, M_s$ are the distinct conjugates of M. Let L be an irreducible submodule of M^G. Then L_H is completely reducible by Clifford's theorem (Theorem 23). Without loss, we may suppose that M_1 is a direct summand of L_H. Then every M_i is a summand of L_H and hence $L_H \cong M_1 \oplus \cdots \oplus M_s$. Thus $\dim_K(L) = \dim_K(M^G)$ and $L = M^G$.

Conversely, if M^G is absolutely irreducible then, by Theorem 35(ii),

$$i_H(M, (M^G)_H) = i_G(M^G, M^G) = 1,$$

the latter equality by Theorem 9, and hence $I(M) = H$ by Theorem 39.

We remark that the converse holds for irreducible induced representations over splitting fields, in particular over $\mathbb{C}$. It might seem as though the condition of absolute irreducibility is required in the second part only for this particular proof. However, the condition cannot be dropped completely as Exercises 10 and 11 will show.

Exercises

1. Let G be a dihedral group of order $2n$, n odd. Show that every two-dimensional irreducible representation of G over $\mathbb{C}$ is induced from a representation of the cyclic subgroup of index 2. Obtain the same result if n is even and $n > 2$. (See Exercises 17 and 18 of Section 2.)
2. Let H be a subgroup of the group G. Suppose that M is a KH-module which affords a permutation representation. Show that M^G affords a permutation representation of G.

3. Let G be a 2-transitive permutation group and let P be the module affording the permutation representation of G over a field K. Show that P has a unique trivial submodule L and a unique maximal submodule M having trivial quotient. Show also that M contains L if and only if the characteristic of K divides the degree of G. Deduce that, if the characteristic of K does not divide the degree of G, then P is the direct sum of the trivial submodule and a nontrivial indecomposable submodule, while if the characteristic of K does divide the degree, then P is indecomposable.
4. Let G be the alternating group A_5. Let H be the normaliser of a Sylow 5-subgroup. Show that $|H| = 10$ and that H is dihedral. Show that G acts 2-transitively on the cosets of H and deduce that G has an irreducible complex representation of degree 5. Let ρ denote the nontrivial one-dimensional (complex) representation of H. Show that ρ^G is the sum of two inequivalent three-dimensional representations of G. Show also that G has a four-dimensional irreducible complex representation.
[Note that $1^2 + 2 \cdot 3^2 + 4^2 + 5^2 = 60$; compare with Corollary 20(iii).]
5. Let G be the symmetric group S_4. Show that the degrees of its irreducible complex representations are $1, 1, 2, 3, 3$. Let ρ be a nontrivial one-dimensional complex representation of the normal subgroup of order 4. Show that ρ has three distinct conjugates. By applying the intertwining number theorem, show that ρ^G is the direct sum of the two three-dimensional irreducible representations of G.
6. Let M and N be KG-modules. Show that $M^* \otimes_K N \cong \mathrm{Hom}_K(M, N)$.
[Compare with Lemma 34.]
7. Let G be a group, H a subgroup of G and K an arbitrary field. Let M be a KH-module. Show that $\mathrm{Hom}_{KH}(KG, M)$ can be given the structure of a KG-module by defining a representation of G on $\mathrm{Hom}_{KH}(KG, M)$ by

$$x \cdot \varphi g = (gx)\varphi$$

whenever $x \in KG$, $g \in G$ and $\varphi \in \mathrm{Hom}_{KH}(KG, M)$.

Let $\{x_1, \ldots, x_n\}$ be a transversal for H in G. Show that, if $g \in G$ and $gx_i = hx_j$, then the formula

$$g[(m \otimes x_i)\psi] = mh \otimes x_j$$

can be extended linearly to define a KG-isomorphism ψ from M^G onto $\mathrm{Hom}_{KH}(KG, M)$.

[Note. The representation defined on $\mathrm{Hom}_{KH}(KG, M)$ is sometimes referred to as *coinduction*. For finite groups, this shows that induction and coinduction are equivalent. See Mackey (1951) where this construction is applied also to infinite groups.]

8. Let G_1 and G_2 be two groups and let K be an arbitrary field. Show that there is a natural identification of $K(G_1 \times G_2)$ with $KG_1 \otimes KG_2$. Let H_1 and H_2 be subgroups of G_1 and G_2 respectively and let M_1 and M_2 be KH_1- and KH_2-modules. Prove that

$$M_1^{G_1} \otimes_K M_2^{G_2} \cong (M_1 \otimes M_2)^{G_1 \times G_2}$$

as $K(G_1 \times G_2)$-modules.
9. (Mackey's tensor product theorem) Let H_1 and H_2 be subgroups of a group G, let K be an arbitrary field, and let M_1 and M_2 be KH_1- and KH_2-modules. Adopting the notation of Theorem 37, show that $M_1^{[x]} \otimes M_2^{[y]}$ depends only on the (H_1, H_2)-double coset to which xy^{-1} belongs, and that

$$M_1^G \otimes M_2^G \cong \bigoplus_{D \in \mathscr{D}} (M_1^{[x]} \otimes M_2^{[y]})^G.$$

[Hint. Consider the restriction of $(M_1 \otimes M_2)^{G \times G}$ to the diagonal subgroup of $G \times G$ and apply Theorem 36.]
10. Let G be cyclic of order 4 and let H be the subgroup of order 2. Show that the nontrivial linear representation of H over any field of characteristic different from 2 is G-stable, but that the induced representation is irreducible if the field does not contain a primitive fourth root of unity.
11. Let G be the quaternion group of order 8 and let H be a subgroup of order 4. Show that a faithful irreducible real representation of H has degree 2 and that, if M is the corresponding module, then M^G is irreducible and $I(M) = H$.
[See also Exercise 11 of Section 5. These two examples show the need for some conditions in the converse part of Corollary 40.]
12. Show that the dihedral group of order 8 does not afford the same 'counterexample' as Exercise 11.
13. Assume the hypothesis of Theorem 39. Show that if $M^{I(M)}$ is irreducible, then so is M^G.

8 Tensor induction and transfer

Let H be a subgroup of a group G and let K be an arbitrary field. In this section, we shall demonstrate how the technique used to define an induced module in Lemma 27 in the last section can be modified to associate a different KG-module with a given module for H. As a particular example, we shall then show how, when the construction is restricted to one-dimensional modules, what emerges is the dual of the transfer homomorphism as defined group-theoretically.

Let $\{g_1, \ldots, g_n\}$ be a transversal for H in G and let M be a KH-module.

Then, instead of taking the direct sum of vector spaces $M \otimes g_i$, we consider their tensor product. As in Theorem 28, we take a K-basis $\{m_1, \ldots, m_t\}$ for M and suppose that M affords a matrix representation ρ with respect to this basis. Now we shall define a vector space

$$M^{\otimes G} = \bigotimes_{i=1}^{n} (M \otimes g_i) \tag{8.1}$$

and show how to give $M^{\otimes G}$ the structure of a KG-module. We remark that, as in the construction of the induced module as a direct sum of spaces $M \otimes g_i$, each such space is isomorphic to M: the notation $\cdot \otimes g_i$ is only for the convenience of the identification of the ith term of a decomposable tensor.

The process described by the theorem below is called *tensor induction*, and the representation of G that the module $M^{\otimes G}$ affords will be denoted by $\rho^{\otimes G}$.

Theorem 41. *On a basis for $M^{\otimes G}$, define*

$$\left[\bigotimes_i (m_{k_i} \otimes g_i) \right] g = \bigotimes_j (m_{k_i} \cdot \rho(g_i g g_j^{-1}) \otimes g_j) \tag{8.2}$$

where, for each i, the index j is taken so that $g_i g g_j^{-1} \in H$, and extend to $\hat{m}g$ by linearity for all $\hat{m} \in M^{\otimes G}$. Then $M^{\otimes G}$ has the structure of a KH-module which is independent (up to isomorphism) of the choice of transversal or basis for M.

Proof. It is a straightforward verification to see first that the action of an element g on $M^{\otimes G}$ is that of a linear transformation, and then that a KG-module has been defined. Furthermore, the formula (8.2) is valid for any decomposable tensor in $M^{\otimes G}$ so that the definition does not depend on the choice of basis for M.

Now suppose that a different transversal had been chosen. If the coset representatives were permuted, then this would correspond only to a permutation of the factors in (8.1) and therefore to a permutation of the elements of the basis taken for $M^{\otimes G}$. In this case, the isomorphism is clear. Thus, we may suppose that a transversal of the form $\{h_1 g_1, \ldots, h_n g_n\}$ has been chosen. Then, since

$$\rho((h_i g_i) g (h_j g_j)^{-1}) = \rho(h_i) \rho(g_i g g_j^{-1}) \rho(h_j)^{-1},$$

the new transversal will give rise to the same matrix representation with respect to a basis for $M^{\otimes G}$ whose members are of the form

$$\bigotimes_i (m_k h_i^{-1}) \otimes g_i.$$

Unlike ordinary induction, tensor induction is not 'linear' but it is multiplicative with respect to taking tensor products.

Theorem 42. *Let L and M be KH-modules. Then*

$$(L \otimes_K M)^{\otimes G} \cong_{KG} (L^{\otimes G}) \otimes_K (M^{\otimes G}).$$

Proof. By taking a transversal for H in G, it is an easy exercise to check that the map

$$(l_1 \otimes m_1) \otimes \cdots \otimes (l_n \otimes m_n) \to (l_1 \otimes \cdots \otimes l_n) \otimes (m_1 \otimes \cdots \otimes m_n)$$

for any $l_1, \ldots, l_n \in L$ and $M_1, \ldots, m_n \in M$ extends to the desired KG-isomorphism.

For the remainder of this section, we shall consider the particular case where ρ is one-dimensional. Then the formula (8.2) reduces to

$$\rho^{\otimes G}: g \to \prod_i \rho(g_i g g_{i_g}^{-1}) \tag{8.3}$$

where $Hg_i g = Hg_{i_g}$ and the representations, being one-dimensional, are identified with scalar multiplications. Now let $V_{G \to H}$ denote the transfer homomorphism from G into H. Then the image $V_{G \to H}(g)$ is given by a formula of the same 'shape'. So we have the following relationship.

Theorem 43. *If ρ is a one-dimensional representation of the subgroup H of G, then $\rho(V_{G \to H}(g)) = \rho^{\otimes G}(g)$.*

It follows that the transfer homomorphism is nontrivial if and only if, for some one-dimensional complex representation ρ of H, tensor induction is nontrivial. (For this observation, it is not necessary to restrict to complex representations: it would be sufficient that the intersection of the kernels of the one-dimensional representations be H'.) In view of Theorem 42, we can make this more precise, recalling that the set of irreducible complex representations of an abelian group form a group (see Exercise 5 of Section 1), and also observing that transfer is really a map from G/G' into H/H'.

Theorem 44. *Let H be a subgroup of the group G and let H^* and G^* be the groups defined by the one-dimensional complex representations of H and G respectively. Then the transfer homomorphism $V_{G \to H}: G/G' \to H/H'$ and the tensor induction map $\cdot^{\otimes G}: H^* \to G^*$ are adjoints.*

Proof. We must be careful by what we mean by adjoints in this context. Let $\tilde{G} = G/G'$ and $\tilde{H} = H/H'$. The groups G^* and H^* are duals of the groups $\tilde{G}$ and $\tilde{H}$ in the sense that they may be viewed as groups of homomorphisms from $\tilde{G}$ and $\tilde{H}$ respectively into the multiplicative group of $\mathbb{C}$. Thus we have maps

$$\tilde{G} \times G^* \to \mathbb{C}^\times$$

and

$$\tilde{H} \times H^* \to \mathbb{C}^\times,$$

and we could regard adjointness merely in the sense of Theorem 43 that

$$(\tilde{g}, \rho^{\otimes G}) = (V_{G \to H}(\tilde{g}), \rho)$$

for all $\tilde{g} \in \tilde{G}$ and $\rho \in H^*$, together with the fact that the KG-isomorphism of Theorem 42 shows that the map $\cdot^{\otimes G}: H^* \to G^*$ is a homomorphism.

However, we can say more. As a vector space, $\mathbb{C}G^*$ is precisely the dual space of $\mathbb{C}\tilde{G}$ where the one-dimensional representations form a basis dual to that consisting of the idempotents of $\mathbb{C}\tilde{G}$, and similarly for $\mathbb{C}H^*$ and $\mathbb{C}\tilde{H}$. Then maps

$$V_{G \to H}: \mathbb{C}\tilde{G} \to \mathbb{C}\tilde{H}$$

and

$$\cdot^{\otimes G}: \mathbb{C}H^* \to \mathbb{C}G^*$$

can be defined as linear extensions, and these are adjoint in the usual contravariant sense of dual spaces. In particular these maps have the same rank so that the observation made after the statement of Theorem 43 may be strengthened to the following.

Corollary 45. $|\mathrm{Im}(\cdot^{\otimes G})| = |V_{G \to H}(g)|$.

So far, this duality may seem a little forced, but we shall see that it is quite natural by using tensor induction to prove the focal subgroup theorem. This was originally proved independently by Brauer (1953) and by D. G. Higman (1953). We shall give Brauer's proof in Chapter 5, while Higman's is the group theoretic equivalent of that given here. We remark that the importance of the focal subgroup theorem lies in its application via Alperin's fusion theorem to obtain many of the classical transfer theorems; we shall give just one example, Burnside's transfer theorem, for which we shall have considerable use later. (See, for example, Gorenstein [G; p. 245ff.].)

Theorem 46 (Focal subgroup theorem). *Let P be a Sylow p-subgroup of the finite group G. Let*

$$P_0 = \langle xy^{-1} \mid x, y \in P, x, y \text{ conjugate in } G \rangle.$$

Then $P \cap G' = P_0$.

Proof. If $y = g^{-1}xg$, then $xy^{-1} = [x^{-1}, g]$ so that $P' \subseteq P_0 \subseteq P \cap G'$. So it remains only to show that $P \cap G' \subseteq P_0$.

Let $\rho \in (P/P_0)^*$, embedding this in $(P/P')^*$ by inflation. Let $x \in P - P_0$. We may choose coset representatives $\{g, gx, gx^2, \ldots\}$ for P in G according to the orbits under $\langle x \rangle$ as for group theoretic transfer, so that the formula (8.3) yields

$$\rho^{\otimes G}(x) = \prod_j \rho(g_j x^{l_j} g_j^{-1})$$

where x permutes the cosets in orbits of length l_j and the representatives $\{g_j\}$ are taken one from each orbit. For each j, we have $g_j x^{l_j} g_j^{-1} \in P$: since $[x^{-l_j}, g_j^{-1}] \in P_0$, we see that

$$\rho(g_j x^{l_j} g_j^{-1}) = \rho(x^{l_j})$$

and hence that

$$\rho^{\otimes G}(x) = \prod_j \rho(x^{l_j}) = \rho(x^{[G:P]}).$$

Since $p \nmid [G:P]$, it follows that $\rho^{\otimes G}$ is trivial if and only if ρ is. Thus the homomorphism $\cdot^{\otimes G}: (P/P_0)^* \to G^*$ is one-to-one and $\mathrm{Im}(\cdot^{\otimes G})$ contains a subgroup of order $[P:P_0]$. By Corollary 45, the same is true for G/G', and hence $P \cap G' = P_0$.

In order to prove Burnside's theorem, we need a lemma which we shall also use a number of times later.

Lemma 47. *Let P be a Sylow subgroup of a group G. If x and y are elements of $Z(P)$ which are conjugate in G, then they are conjugate in $N(P)$.*

Proof. If $y = x^g$, then P and P^g are Sylow subgroups of $C_G(y)$. Hence there exists $h \in C_G(y)$ such that $P = P^{gh}$. But then $y = y^h = x^{gh}$ while $gh \in N(P)$.

Corollary 48 (Burnside's transfer theorem). *Let G be a group and let P be a Sylow p-subgroup. If $P \subseteq Z(N_G(P))$, then G has a normal p-complement.*

Proof. Certainly P is abelian and the hypothesis implies that any element of P is conjugate in $N(P)$ to no other element of P, nor conjugate in G by Lemma 47. By the focal subgroup theorem, $P \cap G' = 1$. Since G/G' is abelian it has a normal p-complement, as then does G.

Another form of transfer, *character theoretic transfer*, has been described by Yoshida (1978). In Exercise 18 of Section 2.3, we shall see that this is precisely the character of the tensor induced representation as defined here.

Exercises

1. Prove an analogue of Mackey's theorem for tensor induction: namely that, with the notation of Theorem 36,
$$(M^{\otimes G})_{H_2} \cong \bigotimes_{i=1}^{m} ((M \otimes x_i)_{H_1^{x_i} \cap H_2})^{\otimes H_2}.$$
2. Let G be the dihedral group of order 8 and let H be the cyclic subgroup of order 4. Let ρ be a faithful irreducible complex representation of H. Compute $\rho^{\otimes G}$.
3. Repeat Exercise 2 for a quaternion group of order 8.
4. (Thompson) Let G be a finite group and let S be a Sylow 2-subgroup of G. Suppose that S has a subgroup S_0 of index 2 and that S contains an element t of order 2 which is conjugate in G to no element of S_0. By letting ρ be the nontrivial representation of S having S_0 in its kernel and computing $\rho^{\otimes G}(t)$, show that G has a subgroup of index 2.
5. Let H be a subgroup of G. Let M be a KH-module affording a permutation representation of H. Show that $M^{\otimes G}$ affords a permutation representation of G.

2

Complex characters

1 Basic properties

Having developed a general theory of representations in the first chapter, we now turn our attention to the representations of finite groups over the field of complex numbers. In view of the results proved previously, it will be clear that the basic properties carry over to any splitting field of characteristic zero (though for some further results it may be necessary that the field be a splitting field for all subgroups too), and it will also be clear that there will be analogues for fields of finite characteristic not dividing the group order. The derivation of such analogues will be left as unspecified exercises since we shall have no use for them.

We fix the following notation. Let G be a finite group and let K be a subfield of $\mathbb{C}$ which is a splitting field for G. Let $\mathscr{C}_1, \ldots, \mathscr{C}_r$ be the conjugacy classes of elements of G and, for $1 \leqq i \leqq r$, define the *class sums* by

$$C_i = \sum_{g \in \mathscr{C}_i} g.$$

By Theorem 1.21, there are exactly r inequivalent irreducible representations $\rho_1, \ldots, \rho_r$ of G over K. Together with Theorem 1.17, this gives a decomposition

$$KG = A_1 \oplus \cdots \oplus A_r \tag{1.1}$$

where $A_1, \ldots, A_r$ are simple algebras, isomorphic to matrix algebras over K, affording the irreducible representations of G (and hence of KG also). In particular, the multiplicity of ρ_i in the right regular representation of G will be equal to its degree (cf. Exercise 8 of Section 1.3). We may now deduce the following result.

Proposition 1. *KG is a semisimple algebra of dimension $|G|$ whose centre $Z(KG)$ has $\{C_1, \ldots, C_r\}$ as a basis. Furthermore,*

(i) *there exist nonnegative integers $\{a_{ijk}\}$ such that, for $1 \leqslant i, j \leqslant r$,*

$$C_i C_j = \sum_{k=1}^{r} a_{ijk} C_k;$$

(ii) *if ρ_i has degree n_i, then $\sum_{i=1}^{r} n_i^2 = |G|$; and*

(iii) *if π is the right regular representation of G, then $\pi \sim \sum_i n_i \rho_i$.*

Proof. KG is semisimple by Maschke's theorem, while the statements about the class sums follow from Theorem 1.21 and the fact that a_{ijk} is the coefficient of any element of the conjugacy class $\mathscr{C}_k$ in the product C_iC_j. Statements (ii) and (iii) come from the decomposition (1.1) above.

Definition. The integers $\{a_{ijk}\}$ of Proposition 1 are called the *class algebra constants* of G.

Any complex representation of G may be realised over K, as was discussed after the proof of Theorem 1.13 in the case of the field $\mathbb{A}$. Thus, for any complex representation ρ of G, the trace values $\operatorname{tr}(\rho(g))$ lie in the field K. So, when considering these values, nothing is lost by studying representations over K.

Definitions. Let ρ be a representation of G over K. The function $\chi: G \to K$ defined by $\chi(g) = \operatorname{tr}(\rho(g))$ is called the (*ordinary* or *complex*) *character* of ρ, and ρ is said to *afford* the character χ. The character χ is said to be *irreducible* if ρ is, *faithful* if ρ is, and have *degree* that of ρ. The *kernel* of χ is the kernel of ρ. If ρ is one-dimensional, then χ is *linear*. The character of the trivial representation is called the *principal character* of G and will be denoted by 1_G.

Proposition 2. (i) *A character is a class function (that is, a function which is constant on conjugacy classes of elements).*

(ii) *Equivalent representations afford the same character.*

Proof. Both statements follow from the fact that, if X is a nonsingular matrix, then $\operatorname{tr}(X^{-1}YX) = \operatorname{tr}(Y)$.

The next result collects together some basic properties of characters. Throughout, we shall use bars to denote complex conjugation in $\mathbb{C}$.

Proposition 3. *Let ρ be a representation of G over K affording a character χ. Then the following hold, where g is any element of G.*

(i) *$\chi(1)$ is equal to the degree of ρ.*
(ii) *$\chi(g)$ is an algebraic integer; in fact, $\chi(g)$ is a sum of nth roots of unity, where n is the order of g.*
(iii) $\chi(g^{-1}) = \overline{\chi(g)}$.
(iv) *$\bar{\chi}$ is the character of G afforded by the contragredient representation ρ^*.*
(v) $|\chi(g)| \leqslant \chi(1)$.

(vi) *If* $|\chi(g)| = \chi(1)$, *then* $\rho(g)$ *is a scalar matrix* (*or scalar transformation*).
(vii) $\chi(g) = \chi(1)$ *if and only if* g *lies in the kernel of* ρ.

Proof. We may assume without loss that $K = \mathbb{C}$. Part (i) is trivial. Suppose that g has order n. Then $(\rho(g))^n = I$. Thus $\rho(g)$ is similar to a diagonal matrix and we may suppose that it is diagonal. Then (ii) and (iii) are clear, while (iv) follows since the matrix form of ρ^* is to take the transpose inverses.

Let $\rho(g)$ have eigenvalues $\lambda_1, \ldots, \lambda_m$. Then

$$\chi(g) = \lambda_1 + \cdots + \lambda_m.$$

So

$$|\chi(g)| = |\lambda_1 + \cdots + \lambda_m| \leqslant |\lambda_1| + \cdots + |\lambda_m| = \chi(1)$$

with equality if and only if $\lambda_1 = \lambda_2 = \cdots = \lambda_m$. The remaining parts of the Proposition follow immediately.

For $i = 1, \ldots, r$, let χ_i denote the character of the irreducible representation ρ_i. In view of the additive nature of traces, the character of an arbitrary representation will be determined by the multiplicities of its irreducible constituents, and by Proposition 2 we may refer to $\chi_1, \ldots, \chi_r$ as *the* irreducible characters of G. We shall now determine the fundamental relations satisfied by these irreducible characters.

Lemma 4. *Let* π *be the character afforded by the right regular representation of* G. *Then*

(i) $\pi(g) = \sum_{i=1}^{r} \chi_i(1)\chi_i(g)$,

(ii) $\pi(1) = \sum_{i=1}^{r} \chi_i(1)^2 = |G|$, *and*

(iii) *if* $g \neq 1$, *then* $\pi(g) = \sum_{i=1}^{r} \chi_i(1)\chi_i(g) = 0$.

Proof. Part (i) follows from Proposition 1(iii). Since $\pi(1) = |G|$ and $\pi(g) = 0$ if $g \neq 1$, (ii) and (iii) follow from (i).

Following the notation of Section 5 of the previous chapter, we shall let $e_1, \ldots, e_r$ be the central primitive idempotents of KG and thus the identity elements of the ideals $A_1, \ldots, A_r$ of the decomposition (1.1) when these ideals are identified with the corresponding matrix algebras. The next result, the explicit computation of these idempotents, is of independent interest.

Theorem 5. $e_i = |G|^{-1} \sum_{g \in G} \chi_i(1)\chi_i(g^{-1})g$.

Proof. Let $e_i = \sum a_g g$. Since traces are additive, we may regard characters as being characters of the group ring KG also. We compute the character π of the regular representation of KG. Then Lemma 4 yields the equation

$$a_g|G| = \pi(e_i g^{-1}) = \sum_{j=1}^{r} \chi_j(1)\chi_j(e_i g^{-1}).$$

Now

$$\rho_j(e_i g^{-1}) = \rho_j(e_i)\cdot\rho_j(g^{-1}) = \begin{cases} 0 & \text{if } i \neq j, \\ \rho_i(g^{-1}) & \text{if } i = j. \end{cases}$$

So $\chi_j(e_i g^{-1}) = \delta_{ij}\chi_i(g^{-1})$ and hence $a_g = |G|^{-1}\chi_i(1)\chi_i(g^{-1})$.

We may now obtain the *orthogonality relations* for ordinary characters. These are of fundamental importance for the entire development of the subject. We fix the following notation: unless otherwise stated, g_i will denote an arbitrary element of the conjugacy class $\mathscr{C}_i$ and we put $h_i = |\mathscr{C}_i|$.

Theorem 6. (i) $|G|^{-1} \sum_{g\in G} \chi_i(g)\overline{\chi_j(g)} = \delta_{ij}$.

(ii) $\sum_{k=1}^{r} \chi_k(g_i)\overline{\chi_k(g_j)} = \delta_{ij}|C_G(g_j)|$.

Proof. We substitute the formula for e_i from Theorem 5 into the product relation

$$e_i e_j = \delta_{ij} e_i,$$

and compare coefficients of the identity element. This yields the equation

$$\frac{\chi_i(1)\chi_j(1)}{|G|^2} \sum_{g\in G} \chi_i((g^{-1})^{-1})\chi_j(g^{-1}) = \frac{\delta_{ij}}{|G|}\chi_i(1)^2.$$

So (i) holds.

With $h_k = |\mathscr{C}_k|$, we may rewrite (i) as

$$|G|^{-1} \sum_{k=1}^{r} h_k \chi_i(g_k)\overline{\chi_j(g_k)} = \delta_{ij}$$

or

$$\sum_{k=1}^{r} \chi_i(g_k)(|C_G(g_k)|^{-1}\overline{\chi_j(g_k)}) = \delta_{ij}.$$

Let A be the $r \times r$ matrix whose (i, k)-entry a_{ik} is $\chi_i(g_k)$ and let B be the $r \times r$ matrix whose (k, j)-entry b_{kj} is $|C_G(g_k)|^{-1}\overline{\chi_j(g_k)}$. Then we have shown that $AB = I$. Thus $BA = I$ and hence, for all i, j,

$$\sum_{k=1}^{r} b_{jk} a_{ki} = \delta_{ij}.$$

This equation may be rewritten as

$$\sum_{k=1}^{r} |C_G(g_j)|^{-1}\overline{\chi_k(g_j)}\chi_k(g_i) = \delta_{ij},$$

from which (ii) follows.

The $r \times r$ matrix A that occurs in this proof is called the *character table* of the group G. The rows are indexed by the irreducible characters of G and the columns by the conjugacy classes. By convention, we take the first row and column to be indexed by 1_G and $\{1\}$ respectively. We note that the second orthogonality relation states that distinct columns of this matrix are orthogonal so that the matrix given by the character table is nonsingular: in particular, the rows of the character table are linearly independent.

We next investigate restrictions on character values and, in particular, on character degrees. After deducing some consequences of the orthogonality relations and finding some further basic properties, we shall construct some examples of character tables; other examples will be given in the exercises at the end of the section.

Theorem 7. (i) $\rho_i(C_j) = ((h_j\chi_i(g_j))/\chi_i(1))I$.
(ii) $(h_j\chi_i(g_j))/\chi_i(1)$ *is an algebraic integer.*

Proof. The class sum C_j lies in the centre of KG and hence $\rho_i(C_j)$ lies in the centre of $\rho_i(KG)$. Now $\rho_i(KG) \cong \mathcal{M}_{n_i}(K)$ and hence $\rho_i(C_j) = \omega_{ij}I$ for some $\omega_{ij} \in K$. On the other hand,

$$\rho_i(C_j) = \sum_{g \in \mathscr{C}_j} \rho_i(g);$$

hence

$$\operatorname{tr}(\rho_i(C_j)) = \sum_{g \in \mathscr{C}_j} \operatorname{tr}(\rho_i(g)),$$

or

$$\omega_{ij}\chi_i(1) = h_j\chi_i(g_j).$$

By Theorem 1.14 and Corollary 1.16, we may suppose that K is an algebraic number field and that, if R is the ring of algebraic integers in K, then ρ_i can be realised in its localisation $R_{\mathscr{P}}$ at any prime ideal $\mathscr{P}$. Since

$$\omega_{ij}I = \sum_{g \in \mathscr{C}_i} \rho_i(g),$$

it follows that $\omega_{ij} \in R_{\mathscr{P}}$ for every prime ideal $\mathscr{P}$ of R. Hence $\omega_{ij} \in R$.

Corollary 8. $\chi_i(1)$ *divides* $|G|$ *for each* i.

Proof. From the first orthogonality relation,

$$\sum_{j=1}^{r} \omega_{ij}\overline{\chi_i(g_j)} = \frac{|G|}{\chi_i(1)}.$$

Hence

$$\frac{|G|}{\chi_i(1)} \in R \cap \mathbb{Q} = \mathbb{Z}.$$

Another consequence of Theorem 7 is a formula for the class algebra constants. This will be a major tool in the applications of character theory that will be discussed in Section 9 and later in this book.

Corollary 9. *The class algebra constants are given by the formulae*

$$a_{ijk} = \frac{|G|}{|C(g_i)|\cdot|C(g_j)|} \sum_{l=1}^{r} \frac{\chi_l(g_i)\chi_l(g_j)\overline{\chi_l(g_k)}}{\chi_l(1)}.$$

Proof. By definition, $C_iC_j = \sum_k a_{ijk}C_k$. Hence, for $l = 1, \ldots, r$,

$$\rho_l(C_i)\cdot\rho_l(C_j) = \sum_k a_{ijk}\rho_l(C_k).$$

By Theorem 7(i), this gives the equation

$$\frac{h_i\chi_l(g_i)}{\chi_l(1)} \cdot \frac{h_j\chi_l(g_j)}{\chi_l(1)} = \sum_{k=1}^{r} a_{ijk} \frac{h_k\chi_l(g_k)}{\chi_l(1)}$$

or, equivalently,

$$\sum_k a_{ijk}h_k\chi_l(g_k) = \frac{h_ih_j}{\chi_l(1)} \chi_l(g_i)\chi_l(g_j).$$

Multiplying both sides by $\overline{\chi_l(g_{k'})}$ and summing over l, we obtain, by applying the second orthogonality relation, the equation

$$a_{ijk'}h_{k'}|C(g_{k'})| = h_ih_j \sum_{l=1}^{r} \frac{\chi_l(g_i)\chi_l(g_j)\overline{\chi_l(g_{k'})}}{\chi_l(1)}$$

and this yields the desired formula for a_{ijk}.

We now obtain two propositions which we shall use in the construction of our examples of character tables. The first is an easy refinement of Proposition 3.

Proposition 10. (i) *If $g^2 = 1$, then $\chi_i(g) \in \mathbb{Z}$.*

(ii) *If g is conjugate to its inverse, then $\chi_i(g) \in \mathbb{R}$.*

(iii) *If g is an element of order 3 which is conjugate to its inverse, then $\chi_i(g) \in \mathbb{Z}$.*

Proof. This follows immediately from the consideration of possible eigenvalues in the matrix representation of a given element.

Next, we derive an important tool for constructing the character tables of doubly transitive permutation groups. We note that the character of a permutation representation is given by the number of fixed points. (See also Exercise 14 in Section 3.)

Proposition 11. *Let G be a group which acts as a doubly transitive permutation group on the set Ω (not necessarily faithfully). Then the character of the permutation representation afforded by the action on Ω is the sum of the principal character and a nonprincipal irreducible character.*

Proof. By Corollary 1.38 (and its application in Exercise 3 of Section 1.7), the permutation representation is the sum of the trivial representation and a nontrivial irreducible representation. Since traces are additive, the result follows.

Examples

1. Let G be the symmetric group S_3. Then G has three conjugacy classes with representatives 1, (12) and (123). By Proposition 1(iii), the degrees of the irreducible characters are $1, 1, 2$. Together with Proposition 10, the second orthogonality relation now determines the character table as follows.

	1	(12)	(123)
χ_1	1	1	1
χ_2	1	-1	1
χ_3	2	0	-1

Notice that these characters may be checked by considering the complex irreducible representations of S_3 that were determined in the exercises in Chapter 1.

2. Let G be the symmetric group S_4. Since S_4 has S_3 as a homomorphic image, this immediately determines three irreducible characters; the sum of the squares of their degrees is 6. There are five conjugacy classes and hence two more irreducible characters, whose degrees must both therefore be 3. The character table may now be completed by means of the second

orthogonality relations alone.

	1	(12)(34)	(12)	(1234)	(123)
$\lvert C(g)\rvert$	24	8	4	4	3
χ_1	1	1	1	1	1
χ_2	1	1	-1	-1	1
χ_3	2	2	0	0	-1
χ_4	3	-1	1	-1	0
χ_5	3	-1	-1	1	0

We remark that the character χ_4 may be written down using Proposition 11, and also that χ_5 is the character afforded by the representation of S_4 as the rotation group of the cube.

3. Let G be the alternating group A_5. Then G has five conjugacy classes, with only the 5-cycles of S_5 splitting into two classes. Recall that A_5 is a simple group and hence has only the trivial character as a linear character. It is easily verified that 59 can be written in only one way as the sum of four squares none of which is 1, and hence the degrees of the irreducible characters are 1, 3, 3, 4, 5. Notice that one of the three-dimensional characters must be that afforded by the rotation group of the regular icosahedron, and the four-dimensional representation may be obtained from the standard permutation representation. Using Proposition 10 in addition, it may then be verified that the character table of A_5 is as follows, where $\alpha = \omega + \omega^{-1}$ and $\beta = \omega^2 + \omega^{-2}$ for a fixed primitive fifth root of unity ω.

	1	(12)(34)	(123)	(12345)	(13524)
$\lvert C(g)\rvert$	60	4	3	5	5
χ_1	1	1	1	1	1
χ_2	3	-1	0	$-\alpha$	$-\beta$
χ_3	3	-1	0	$-\beta$	$-\alpha$
χ_4	4	0	1	-1	-1
χ_5	5	1	-1	0	0

We also observe that χ_3 is afforded by the representation obtained as the composite of the automorphism induced by conjugation by a transposition in S_5 (which interchanges the conjugacy classes of elements of order 5) with the representation affording χ_2.

We now turn to some slightly deeper properties of character tables which can assist in their construction. The following results will be used extensively later in this book. The first is due to Brauer.

Theorem 12. *Let σ act as a permutation of each of the sets of irreducible characters of a group G and the elements of G. Suppose that σ preserves conjugacy of elements in G and that*

$$\chi^\sigma(g) = \chi(g^\sigma),$$

where χ^σ and g^σ denote the images of χ and g under σ respectively. Then the number of fixed characters and the number of fixed conjugacy classes under σ are the same.

Proof. Let X be the matrix $(\chi_i(g_j))$, namely the character table. Since the columns of X are nonzero and mutually orthogonal, X is nonsingular. Let P_1 and P_2 be the permutation matrices corresponding to permutations induced by σ on the rows and columns of X. Then the hypothesis of the theorem may be restated as a matrix equation

$$P_1 X = X P_2,$$

or

$$X^{-1} P_1 X = P_2.$$

Thus P_1 and P_2 have the same trace, so that the numbers of fixed points are the same.

We note two special cases.

(a) If σ is an automorphism of G, then we can *define* χ^σ by the formula

$$\chi^\sigma(g) = \chi(g^\sigma).$$

This is precisely the situation that occurred with A_5 above.

(b) Let a permutation σ be defined on characters by $\chi^\sigma = \bar{\chi}$ and on G by $g^\sigma = g^{-1}$.

Definition. (i) A character χ of G is said to be *real* (resp., *rational*) if $\chi(g)$ is real (resp., rational) for all $g \in G$.

(ii) An element g of G is *real* if g and g^{-1} are conjugate.

Applying Theorem 12 with the map σ given in case (b) above, we immediately deduce the following relationship between these definitions.

Proposition 13. *The number of complex irreducible characters of a group G that are real is equal to the number of conjugacy classes of real elements. In particular,*

(i) *an element g is real if and only if $\chi(g)$ is real for every irreducible character χ, and*
(ii) *if G has odd order, then only the principal character is real.*

We remark that a complex representation may have a real character, yet not be realisable over $\mathbb{R}$. This will be discussed further in Section 8.

We now examine a more general situation than complex conjugation. The next result is an immediate consequence of the construction of Example 4 of Section 1.2 where field automorphisms were used to construct new representations.

Proposition 14. *Let K be an algebraic number field which is a splitting field for the group G. Let α be an automorphism of K. If χ is a character of G, define a function χ^α on G by*

$$\chi^\alpha(g) = (\chi(g))^\alpha.$$

Then χ^α is a character of G which is irreducible if and only if χ is.

This proposition leads immediately to the following definition. The general theory of field extensions ensures that the definition does not depend on the particular field K chosen.

Definition. Let K be a finite normal extension of $\mathbb{Q}$ which is a splitting field for the group G. Let χ and χ' be characters of G. If there exists an automorphism α of K such that $\chi' = \chi^\alpha$, then χ and χ' are *algebraically conjugate.*

As an example, we note that the two three-dimensional characters of A_5 are algebraically conjugate under the automorphism $\omega \to \omega^2$ of $\mathbb{Q}(\omega)$, where ω is a primitive fifth root of unity. However, the two three-dimensional representations of S_4 are *not* algebraically conjugate.

Suppose now that ε is a primitive $|G|$th root of unity and that K is a normal extension of $\mathbb{Q}$ containing $\mathbb{Q}(\varepsilon)$. Then, if $(m, |G|) = 1$, there exists an automorphism α_m of K such that $\varepsilon^{\alpha_m} = \varepsilon^m$.

Lemma 15. *If χ is a character of G, define a function χ^{α_m} on G by*

$$\chi^{\alpha_m}(g) = \chi(g^m).$$

Then χ^{α_m} is a character of G which is irreducible if and only if χ is.

Proof. We may suppose, first, that the splitting field K for G is a normal extension of $\mathbb{Q}$ containing $\mathbb{Q}(\varepsilon)$. Since $\chi(g)$ is a sum of powers of ε, it follows that

$$\chi(g^m) = (\chi(g))^{\alpha_m}$$

and Proposition 14 gives χ^{α_m} as an algebraic conjugate of χ. But now that χ^{α_m} is a character, we know that the map α_m can be defined with reference to the field $\mathbb{Q}(\varepsilon)$ only, so the result follows.

We complete this section with a brief discussion of rational characters.

Theorem 16. *Let G be a group in which, whenever $(m, |G|) = 1$, the elements g* and g^m *are conjugate for all $g \in G$. Then every irreducible character of G is rational.*

Proof. Without loss, we may take a splitting field K which is a normal extension of $\mathbb{Q}$ containing a primitive $|G|$th root ε of unity. The Galois group of $\mathbb{Q}(\varepsilon)$ over $\mathbb{Q}$ consists of all automorphisms defined by $\alpha_m : \varepsilon \to \varepsilon^m$ where $(m, |G|) = 1$. By hypothesis, $\chi^{\alpha_m} = \chi$ for all such m; hence $\chi_i(g) \in \mathbb{Q}$ for all $g \in G$.

As a particular case, the hypothesis of Theorem 16 holds for the symmetric groups. Thus we have the following consequence.

Corollary 17. *The characters of the symmetric groups are rational.*

Exercises

1. (i) Let ρ_m and ρ_n be complex irreducible matrix representations of a group G and suppose that $\rho_m(g) = (a_{ij}(g))$ and $\rho_n(g) = (b_{ij}(g))$. For a matrix S of appropriate size, define

$$f(S) = \sum_{g \in G} \rho_m(g^{-1}) S \rho_n(g).$$

Show that $f(S)$ intertwines ρ_m and ρ_n.

(ii) Let E_{st} be the matrix with (s, t)-entry 1 and all other entries zero. By considering the (i, j)-entry of $f(E_{st})$, show that, if ρ_m and ρ_n are inequivalent, then

$$\sum_{g \in G} a_{is}(g^{-1}) b_{tj}(g) = 0$$

for all i, j, s, t.

(iii) If $\rho_m = \rho_n$, use Schur's Lemma to show that

$$\sum_{g \in G} a_{is}(g^{-1}) a_{tj}(g) = \frac{|G|}{\deg(\rho_m)} \delta_{ij} \delta_{st}$$

for all i, j, s, t.

[Note. These are known as the *Schur relations*.]

2. Deduce from Exercise 1(iii) above, along the lines of the proof of Theorem 7(ii), that $\deg(\rho_m)$ divides $|G|$.
3. Refine the argument of Exercise 2 to show that $\deg(\rho_m)$ divides the index $[G:Z(G)]$.
4. Use the Schur relations to prove the first orthogonality relation directly.
5. Derive Theorem 5 (the formula for the primitive central idempotents) from the orthogonality relations.
6. Show that normal subgroups may be identified from the character table of a group. What is the index of the derived group?
7. Show that, if a group G has a faithful irreducible character, then $Z(G)$ is cyclic. Show that the converse holds for p-groups.
8. Determine the character table of the alternating group A_4.
9. Determine the character tables of the dihedral groups.
   [See Exercises 8 and 9 in Section 1.1 for a description of the irreducible complex representations.]
10. Determine the character tables of the generalised quaternion groups.
   [See Exercise 10 of Section 1.1. Note that, if G is a generalised quaternion group, then $G/Z(G)$ is dihedral. Also note that the groups D_8 and Q_8 have the same character tables. However, their faithful irreducible *real* representations have degrees 2 and 4 respectively; see Exercise 11 of Section 1.5 for Q_8, and consider the action of D_8 on a square in $\mathbb{R}^2$. See also Section 8 (in particular, Exercise 3).]
11. If χ is the character of a group G afforded by a representation ρ, show that the function $(\det \chi)$ defined by

$$(\det \chi)(g) = \det(\rho(g))$$

   is a linear character of G, called a *determinantal character*. Show also that the nonlinear irreducible characters of D_8 and Q_8 determine different determinantal characters (so that determinantal characters are not determined by the character table of a group).
12. Determine the character tables of the symmetric groups S_5 and S_6.
13. Determine the character table of the group $\mathrm{SL}(2, 5)$.
   [Note. $\mathrm{PSL}(2, 5) \cong A_5$.]

14. Determine the character table of A_6.
[Note. $A_6 \cong \mathrm{PSL}(2,9)$. This acts as a doubly transitive permutation group on 10 letters.]
15. Determine the character table of the group $\mathrm{PSL}(2,7)$.
[Note. $\mathrm{PSL}(2,7) \cong \mathrm{GL}(3,2)$. Thus the conjugacy classes can be described by rational canonical forms. Also, there are two natural doubly transitive permutation representations.]
16. Let $z \in Z(G) \cap G'$, where $Z(G)$ denotes the centre of a group G. Suppose that $z^2 = 1$ and that χ is an irreducible character of G such that $z \notin \ker \chi$. Show that χ has even degree.
17. Let G be a group and let H be a normal subgroup of G. Suppose that G contains an element x such that $C_G(x) \cap H = 1$. Show that $\chi(x) = 0$ for every irreducible character χ of G for which $H \nsubseteq \ker \chi$.
18. Let G be a nonabelian group of order pq, where p and q are primes with $p < q$. Show that G has a normal Sylow q-subgroup Q and that every element outside Q has order p. Show that the number of conjugacy classes in G is $p + (q-1)/p$, and deduce that every nonlinear irreducible character has degree p. Find the complete character table.
19. Let G be a simple group having a Hall subgroup H of odd order. Assume that G has a nonlinear irreducible character χ whose degree divides $|H|$. Show that no involution τ of G can centralise H.
[Hint. Note that $|\chi(\tau)| < \chi(1)$. Use Theorem 7(ii). Recall that a *Hall subgroup* of G is a subgroup whose order is coprime to its index.]
20. Show that the irreducible characters are linearly independent in the space of complex-valued functions on G.
21. Let φ and ψ be characters of a group G afforded by representations ρ and σ respectively. Show that the characters afforded by $\rho + \sigma$ and $\rho \otimes \sigma$ are $\varphi + \psi$ and $\varphi\psi$.
22. Let A be a group which acts as a permutation group both on the set of irreducible characters of a group G and on the elements of G, preserving conjugacy. Suppose that $\chi^a(g) = \chi(g^a)$ for all irreducible characters χ and all $a \in A$ and $g \in G$. Show that the numbers of orbits of A on the set of irreducible characters and on the set of conjugacy classes are the same.
23. Let H be a subgroup of a group G. Show that the restriction of a character of G to H is a character of H. Show that, if the restriction is irreducible, then so is the original character of G, but that the converse need not hold.
24. Let χ be a character of the group G. Show that the product of the algebraic conjugates of χ is a rational character.
25. Let χ be a character of a cyclic group G. Show that $\prod_X \chi(g)$ is rational,

where $X = \{g \in G \mid G = \langle g \rangle\}$. Deduce that, if $\chi(g) \neq 0$, then

$$\prod_X |\chi(g)|^2 \geqslant 1$$

and hence that

$$\sum_X |\chi(g)|^2 \geqslant |X|.$$

26. (Burnside) Let G be a group and let χ be a nonlinear irreducible character of G. Show that $\chi(g) = 0$ for some $g \in G$.
[Hint. First show that an equivalence relation on the elements of G can be defined by $g \sim h$ whenever $\langle g \rangle = \langle h \rangle$. Then apply Exercise 25.]
27. Translate Clifford's theorem and the associated results of Section 1.6 into results about characters.
28. Show that the only algebra homomorphisms from $Z(KG)$ into K are given by the functions

$$\delta_i : e_j \to \delta_{ij}.$$

Show that these are precisely the functions

$$\omega_i : C_j \to h_j \chi_i(g_j)/\chi_i(1)$$

that arise out of Theorem 7.
29. (G. Higman) Let G be an abelian group and let $e_1, \ldots, e_r$ be the central primitive idempotents of KG.

(i) Show that $e_1, \ldots, e_r$ form a basis for KG.
(ii) Let $u \in KG$, and suppose that $u = \sum a_g g = \sum b_i e_i$. Show that

$$a_g = |G|^{-1} \sum_i b_i \chi_i(g) \quad \text{and} \quad b_i = \sum_g a_g \chi_i(g).$$

(iii) Suppose that $u \in \mathbb{Z}G$, with $u^m = 1$ for some $m \geqslant 1$. Show that $|b_i| \leqslant 1$ and hence that $|a_g| \leqslant 1$. Deduce, using the orthogonality relations, that $u = \pm g$ for some $g \in G$.

30. Show that an element in a group is rational if and only if every character takes rational values on it.
[An element g in a group G is *rational* if g is conjugate to g^n whenever n is coprime to the order of g.]

2 Burnside's $p^a q^b$-theorem

We shall devote this section to the proof of the following classical result of Burnside.

Theorem 18. *A finite group whose order is divisible by at most two distinct primes is soluble.*

Since a p-group is certainly soluble, we may assume that the order of a group under consideration is of the form $p^a q^b$ where p and q are distinct primes and a and b are positive integers. For this reason, the theorem is usually called the $p^a q^b$-theorem. For many years, only a character theoretic proof was known, but a fairly short group theoretic proof has now been given by Goldschmidt (1970) and Matsuyama (1973) using some of the more modern ideas in group theory. (See also Suzuki [S II; p. 216].) However, in view of the powerful application that it gives of character theoretic methods, we include a proof here; indeed, it should be noted that no proof free of character theory has yet been given for the critical intermediate result, Theorem 21.

We start the proof of Theorem 18 with a number theoretic lemma.

Lemma 19. *If α is a sum of n roots of unity and α/n is an algebraic integer, then either $\alpha/n = 0$ or α/n is a root of unity.*

Proof. Let α/n have minimal polynomial $f(x)$ over $\mathbb{Q}$. Then $f(x)$ is monic and integral. In particular, the constant term c is an integer so that, if $\beta_1, \ldots, \beta_t$ and the conjugates of α/n, then $\prod_i \beta_i$ is an integer.

If $\alpha/n = (\lambda_1 + \cdots + \lambda_n)/n$ with each λ_j a root of unity, then each β_i is such a sum also. In particular, $|\beta_i| \leqslant 1$. If any $|\beta_i| < 1$, then $|c| < 1$ so that $c = 0$ and $\alpha/n = 0$. Otherwise $|\alpha/n| = 1$ and $\alpha = n\lambda$.

The following lemma and theorem are general results, and we assume the notation of the previous section.

Lemma 20. *Suppose that $(\chi_i(1), h_j) = 1$ for some i, j. Then either $\chi_i(g_j) = 0$ or $\chi_i(g_j) = \lambda\chi_i(1)$ for some root λ of unity.*

Proof. There exist integers a and b such that $a\chi_i(1) + bh_j = 1$. So

$$a\chi_i(g_j) + bh_j\frac{\chi_i(g_j)}{\chi_i(1)} = \frac{\chi_i(g_j)}{\chi_i(1)}.$$

Applying Theorem 7(ii) to the second term, it follows that $\chi_i(g_j)/\chi_i(1)$ is an algebraic integer. Now Lemma 19 applies.

Theorem 21. *Let g be an element of the group G and suppose that g has p^a conjugates, where p is prime and $a \geqslant 1$. Then G is not simple.*

Proof. From the second orthogonality relation,

$$1 + \sum \chi(1)\chi(g) = 0$$

where the sum is taken over all nonprincipal irreducible characters χ of G. If p were to divide $\chi(1)$ whenever $\chi(g) \neq 0$, then p^{-1} would be an algebraic integer, which it is not. So $\chi(g) = \lambda\chi(1)$ for some nonprincipal character χ, by Lemma 20. Thus $\rho(g) \in Z(\rho(G))$ where ρ is the representation affording χ, and G cannot be simple.

We may now prove Burnside's theorem. Suppose that X is a noncyclic composition factor in a group of order $p^a q^b$. Let g be a nonidentity element in the centre of a Sylow subgroup of X. Then Theorem 21 yields a contradiction.

Exercises

1. Let G be a group having a nonlinear faithful irreducible character χ of degree p^a, p prime. Suppose that $Z(G) = 1$. Show that χ vanishes on the nonidentity elements in the centre of a Sylow p-subgroup of G.
2. Show that a nonabelian simple group cannot have a nilpotent subgroup of prime power index.

3 The character ring: restriction and induction

In this section, we begin the study of the characters of a group and its subgroups viewed as families of class functions. We continue to use the general notation established in Section 1.

Let $\mathscr{C}(G, \mathbb{C})$ denote the $\mathbb{C}$-algebra of complex-valued class functions on a group G. Regarding $\mathscr{C}(G, \mathbb{C})$ as a vector space, define, for $\varphi, \psi \in \mathscr{C}(G, \mathbb{C})$, a form $(\cdot, \cdot)$ by

$$(\varphi, \psi)_G = |G|^{-1} \sum_{g \in G} \varphi(g)\overline{\psi(g)},$$

where we will suppress the suffix if there is no risk of ambiguity, and define a norm $\|\cdot\|$ by

$$\|\varphi\|_G^2 = (\varphi, \varphi)_G.$$

This form has an important property when restricted to characters.

Lemma 22. *The form $(\cdot, \cdot)$ is symmetric on characters.*

Proof. We may simply compute that, for characters φ, ψ,

$$\begin{aligned}(\varphi, \psi) &= |G|^{-1} \sum \varphi(g)\overline{\psi(g)} \\ &= |G|^{-1} \sum \varphi(g)\psi(g^{-1}) \\ &= |G|^{-1} \sum \psi(g)\varphi(g^{-1}) \\ &= |G|^{-1} \sum \psi(g)\overline{\varphi(g)} = (\psi, \varphi).\end{aligned}$$

The next result is fundamental for the subsequent development of character theory.

Theorem 23. *The form* $(\cdot,\cdot)$ *is a positive definite Hermitian inner product on* $\mathscr{C}(G,\mathbb{C})$. *With respect to this product, the irreducible characters* $\chi_1,\ldots,\chi_r$ *form an orthonormal basis.*

Proof. The first statement is a trivial verification. The orthogonality relations for characters imply that $(\chi_i,\chi_j)=\delta_{ij}$ and, since $\dim_{\mathbb{C}}\mathscr{C}(G,\mathbb{C})=r$, it follows that $\chi_1,\ldots,\chi_r$ form an orthonormal basis for $\mathscr{C}(G,\mathbb{C})$.

We draw a number of conclusions, for some of which we leave the proofs as exercises.

Corollary 24. *If* χ *is a character of* G, *there exist unique nonnegative integers* $a_1,\ldots,a_r$ *such that* $\chi=\sum a_i\chi_i$, *and* $a_i=(\chi,\chi_i)_G$.

Proof. If ρ is a representation affording the character χ, then Maschke's theorem implies that there exist nonnegative integers $a_1,\ldots,a_r$ such that $\rho\sim\sum a_i\rho_i$. The uniqueness and explicit values now follow from the fact that $\chi_1,\ldots,\chi_r$ form an orthonormal basis.

Following the terminology for representations, we call the integer a_i the *multiplicity* of χ_i as a *constituent* of χ. If every $a_i\leqslant 1$, we say that χ is *multiplicity-free.*

Corollary 25. *Two complex representations of* G *are equivalent if and only if they afford the same character.*

Proof. Equivalent representations afford the same character by Proposition 2. Conversely, if ρ and ρ' afford the same character χ, then the multiplicity of ρ_i as a constituent of each must be (χ,χ_i) so that ρ and ρ' are similar.

Corollary 26. *If* χ *is a character, then* χ *is irreducible if and only if* $\|\chi\|=1$.

Corollary 27. *A nonzero class function* φ *is a character if and only if* (φ,χ_i) *is a nonnegative integer for all* i.

Since $\mathscr{C}(G,\mathbb{C})$ is a ring, the sum and product of characters are certainly class functions. However, it is easily verified that they are in fact the characters afforded by the sum and tensor product of their corresponding representations (Exercise 21 of Section 1). Consequently, we may make the following definition.

Definition. The $\mathbb{Z}$-submodule of $\mathscr{C}(G,\mathbb{C})$ spanned by the irreducible characters of G is called the *character ring* of G. It will be denoted by $\mathscr{C}\mathscr{h}(G)$. The elements of $\mathscr{C}\mathscr{h}(G)$ are called *generalised characters.*

In view of the preceding remarks, $\mathscr{C}\mathscr{h}(G)$ is a ring and any generalised character is the difference of two characters. For this reason, they are sometimes called difference characters. We note that, by Corollary 24, the irreducible characters $\chi_1,\ldots,\chi_r$ form a $\mathbb{Z}$-basis for $\mathscr{C}\mathscr{h}(G)$ (as a free $\mathbb{Z}$-module). In much of the work in subsequent chapters, we shall not be able to distinguish between a character and its negative; specifically, our fundamental objects will be generalised characters of norm 1 rather than irreducible characters. For this reason, we shall allow the term multiplicity to include negative possibilities, and by the term *multiplicity-free*, we shall mean that the multiplicities of the irreducible constituents of a generalised character have absolute value at most 1.

Now let H be a subgroup of G and suppose that $\varphi\in\mathscr{C}(G,\mathbb{C})$. Then the restriction $\varphi|_H$ of φ to H is certainly constant on conjugacy classes of H, and it is easy to check that the restriction map $\varphi\to\varphi|_H$ defines a linear transformation

$$|_H:\mathscr{C}(G,\mathbb{C})\to\mathscr{C}(H,\mathbb{C})$$

and that the restriction of this map to $\mathscr{C}\mathscr{h}(G)$ yields a $\mathbb{Z}$-homomorphism into $\mathscr{C}\mathscr{h}(H)$. Now, since the spaces $\mathscr{C}(G,\mathbb{C})$ and $\mathscr{C}(H,\mathbb{C})$ are equipped with nonsingular Hermitian forms, there is a unique adjoint transformation

$$*:\mathscr{C}(H,\mathbb{C})\to\mathscr{C}(G,\mathbb{C})$$

such that, if $\theta\in\mathscr{C}(H,\mathbb{C})$, then $\theta\to\theta^*$ and, for all $\theta\in\mathscr{C}(H,\mathbb{C})$ and $\varphi\in\mathscr{C}(G,\mathbb{C})$,

$$(\theta,\varphi|_H)_H=(\theta^*,\varphi)_G.$$

Definition. Let H be a subgroup of G and suppose that $\theta\in\mathscr{C}(G,\mathbb{C})$. Then the function θ^* is called the *induced function* and will be denoted by θ^G.

Our first result shows that the process of induction is transitive.

Theorem 28. *Suppose that $H\subseteq L\subseteq G$. If $\theta\in\mathscr{C}(H,\mathbb{C})$, then $\theta^G=(\theta^L)^G$.*

Proof. θ^G is the unique element of $\mathscr{C}(G, \mathbb{C})$ such that

$$(\theta^G, \varphi)_G = (\theta, \varphi|_H)_H$$

for all $\varphi \in \mathscr{C}(G, \mathbb{C})$. On the other hand, for all such functions φ,

$$((\theta^L)^G, \varphi)_G = (\theta^L, \varphi|_L)_L = (\theta, \varphi|_H)_H.$$

The importance of our definitions for character theory is given by the following theorem, only the second part of which still requires proof.

Theorem 29. *Suppose that H is a subgroup of G. Let θ be a character of H and let χ be a character of G. Then the following hold.*

(i) $\chi|_H$ *is a character of H.*
(ii) θ^G *is a character of G.*
(iii) $(\theta, \chi|_H)_H = (\theta^G, \chi)_G$.

Proof. Suppose that, in our standard notation, $\theta^G = \sum_i a_i \chi_i$ where $a_1, \ldots, a_r \in \mathbb{C}$. Then

$$a_i = (\theta^G, \chi_i)_G = (\theta, \chi_i|_H)_H.$$

Now θ and $\chi_i|_H$ are nonnegative integral linear combinations of the irreducible characters of H. Hence a_i is a nonnegative integer and so θ^G is a character of G by Corollary 27.

In view of the linearity of the maps given by restriction and induction, we may conclude that the statements of this theorem have exact analogues for generalised characters. We shall show that our definition of induced characters gives precisely the same class functions as would the characters of induced representations; then our definition via adjoints coincides with the semisimple case of the Nakayama relations, a formula which was established by Frobenius. Thus Part (iii) of Theorem 29 is usually called the *Frobenius reciprocity theorem*.

In order to establish this claim, and also to be able to perform actual computations of characters, we next determine the values of induced class functions.

Theorem 30. *Let H be a subgroup of G and suppose that $\theta \in \mathscr{C}(H, \mathbb{C})$. Then, if $g \in G$,*

$$\theta^G(g) = |H|^{-1} \sum_{x \in G} \theta_0(xgx^{-1})$$

where

$$\theta_0(y) = \begin{cases} 0 & \text{if } y \notin H, \\ \theta(y) & \text{if } y \in H. \end{cases}$$

(Note. We take the conjugate xgx^{-1}, rather than $x^{-1}gx$, to establish consistency with our work in Chapter 1 on induced representations.)

Proof. We have two linear maps from $\mathscr{C}(H,\mathbb{C})$ to $\mathscr{C}(G,\mathbb{C})$, namely

$$\theta \to \theta^G$$

and

$$\theta \to \theta' = |H|^{-1} \sum_{x \in G} \theta_0(x \cdot x^{-1}).$$

So it is sufficient to show that they agree on a basis of $\mathscr{C}(H,\mathbb{C})$. For each conjugacy class $\mathscr{D}$ of H, let $\psi_{\mathscr{D}}$ be the characteristic function. Then the collection $\{\psi_{\mathscr{D}}\}$ forms a basis for $\mathscr{C}(H,\mathbb{C})$.

Let $\varphi_{\mathscr{C}}$ be the characteristic function of the conjugacy class $\mathscr{C}$ in G, and let $g \in \mathscr{C}$. Then

$$(\psi_{\mathscr{D}}, \varphi_{\mathscr{C}}|_H)_H = |H|^{-1}\delta|\mathscr{D}|$$

where $\delta = 1$ if $\mathscr{D} \subseteq \mathscr{C}$ and $\delta = 0$ otherwise. On the other hand,

$$\psi'_{\mathscr{D}}(g) = \delta \cdot |H|^{-1} \cdot |C_G(g)| \cdot |\mathscr{D}|.$$

Thus

$$\begin{aligned}(\psi'_{\mathscr{D}}, \varphi_{\mathscr{C}})_G &= |G|^{-1} \cdot |\mathscr{C}| \cdot \delta \cdot |H|^{-1} \cdot |C_G(g)| \cdot |\mathscr{D}| \\ &= \delta \cdot |H|^{-1} \cdot |\mathscr{D}| = (\psi_{\mathscr{D}}, \varphi_{\mathscr{C}}|_H)_H.\end{aligned}$$

Hence, since the functions $\{\varphi_{\mathscr{C}}\}$ form a basis for $\mathscr{C}(G,\mathbb{C})$ and adjoints are unique, $\psi'_{\mathscr{D}} = \psi^G_{\mathscr{D}}$.

Corollary 31. *Let $\{g_1, \dots, g_n\}$ be a (right) transversal for H in G. Then*

$$\theta^G(g) = \sum_{i=1}^{n} \theta_0(g_i g g_i^{-1}).$$

Proof. For $h \in H$, we have $\theta_0((hg_i)g(hg_i)^{-1}) = \theta_0(g_i g g_i^{-1})$. So the result is clear.

Corollary 32. *Let ρ be a representation of H affording the character θ. Then the induced representation ρ^G affords the character θ^G of G.*

Proof. We consider explicit matrix representations for ρ and ρ^G as described by Theorem 1.28. For $g \in G$, the trace of the matrix $\rho^G(g)$ is given by

$$\operatorname{tr}(\rho^G(g)) = \sum_{i=1}^{n} \operatorname{tr} R_{ii}(g) = \sum{}' \operatorname{tr} \rho(g_i g g_i^{-1})$$

where the second summation is carried out over those i for which $g_i g g_i^{-1} \in H$. But this gives $\theta^G(g)$ by Corollary 31.

Exercises

1. Establish Corollaries 26 and 27.
2. Let K be a subfield of $\mathbb{C}$ which is a splitting field for G. Let M and N be KG-modules affording characters χ and φ respectively. Show that $i(M,N)=(\chi,\varphi)=(1_G,\bar{\chi}\varphi)$.
[Compare this with Lemma 1.34.]
3. Let χ be a character of G. Show that the multiplicity of 1_G in $\chi\bar{\chi}$ is $\|\chi\|^2$. Deduce that if χ is a nonlinear irreducible character, then $\chi\bar{\chi}$ is reducible.
4. Let K be as in Exercise 2 and let M be an irreducible KG-module. Show that $M\otimes M$ contains the trivial module with multiplicity 1 or 0 according to whether M is isomorphic to its dual or not.
5. Let G and H be groups with sets of irreducible characters $\{\varphi_i\}$ and $\{\psi_j\}$. Show that the irreducible characters of $G\times H$ are given by the functions $\{\theta_{ij}\}$ where

$$\theta_{ij}(g,h)=\varphi_i(g)\psi_j(h).$$

6. Let P be a Sylow p-subgroup of G and suppose that χ is a character of G such that $\chi(g)=0$ for all $g\in P-\{1\}$. By considering $\chi|_P$, show that $\chi(1)\equiv 0(\mathrm{mod}\,|P|)$.
7. Let H be a subgroup of G and let φ be a character of H. Show that $\varphi^G(1)=[G:H]\varphi(1)$.
8. Let N be a normal subgroup of G. If φ is a character of N, show that φ^G vanishes outside N.
9. Show that every nonlinear irreducible character of a dihedral group is induced from an irreducible character of the cyclic subgroup of index 2.
10. Let θ be a generalised character of a subgroup H of G. Show that 1_G is a constituent of θ^G if and only if 1_H is a constituent of θ.
11. Show that the irreducible character of degree 5 of the alternating group A_5 is induced from a character of a subgroup.
12. Let G act as a transitive group of permutations on a set Ω with permutation character π. Let H be the stabiliser of a point in Ω. Show that $\pi=(1_H)^G$.
13. Let G act as a group of permutations on a set Ω with permutation character π. Show that the number of orbits is $(1,\pi)$.
14. Let G act as a transitive group of permutations on a set Ω with permutation character π. By considering the action of G on $\Omega\times\Omega$, show that the rank of G is $\|\pi\|^2$.
[Compare this with Proposition 11 in the case that G acts doubly transitively.]

15. (Mackey's Decomposition Theorem) Let H_1 and H_2 be subgroups of a group G and let $\{x_1, \dots, x_m\}$ be a set of (H_1, H_2)-double coset representatives in G. Let $\theta \in \mathscr{C}(H_1, \mathbb{C})$ and, for $i = 1, \dots, m$, put
$$\theta_i(g) = \theta(x_i g x_i^{-1}) \quad \text{for } g \in H_1^{x_i} \cap H_2.$$
Prove, from the definitions of induced class functions, that
$$\theta^G|_{H_2} = \sum_{i=1}^{m} (\theta_i|_{H_1^{x_i} \cap H_2})^{H_2}.$$
If $\varphi \in \mathscr{C}(H_2, \mathbb{C})$, show that
$$(\theta^G, \varphi^G) = \sum_{i=1}^{m} (\theta_i, \varphi)_{H_1^{x_i} \cap H_2}$$
where, on the right-hand side, the obvious restrictions are implied. [Compare with the direct translation of Theorem 1.36.]
16. Establish Exercise 14 by application of Exercise 15.
17. (Brauer) Let G be a simple group and let H_1 and H_2 be subgroups. Suppose that χ is a nonprincipal irreducible character of G such that
$$(1, \chi|_{H_1 \cap H_2}) < (1, \chi|_{H_1}) + (1, \chi|_{H_2}).$$
Show that $\langle H_1, H_2 \rangle$ is a proper subgroup of G.
18. Show that the map $\chi \to \det \chi$ can be extended to define a homomorphism from the additive group of $\mathscr{C}h(G)$ to the multiplicative group of $\mathbb{C}$. Let H be a subgroup of G and define a map $T^G: \mathscr{C}h(H) \to \mathscr{C}h(G)$ by
$$T^G(\theta) = \det(\theta^G - (1_H)^G).$$
Show that $T^G(\theta) = T^G(\det \theta)$ and that, if θ is the character of a one-dimensional representation ρ of H, then $T^G(\theta)$ is the character of $\rho^{\otimes G}$.
[Note. See Exercise 11 of Section 1 for the definition of $\det \chi$. The formula above is the definition of *character theoretic* transfer as given by Yoshida (1978). See also Section 1.8.]
19. Let H be a subgroup of a group G and let $\theta \in \mathscr{C}(G, \mathbb{C})$ and $\varphi \in \mathscr{C}(H, \mathbb{C})$. Prove, by direct computation, that $\theta \cdot \varphi^G = (\theta|_H \cdot \varphi)^G$.
[Compare this with Proposition 1.32 in the case that $\theta \in \mathscr{C}h(G)$ and $\varphi \in \mathscr{C}h(H)$.]

4 Frobenius' theorem

We now turn to the second of the classical results proved by means of character theory. In this case, however, even today there is no proof known in which character theory is not required, and the basic ideas of the proof lie behind much of the work of the later chapters. We start with the permutation theoretic formulation.

Theorem 33 (Frobenius). *Let G be a transitive permutation group of degree n acting on a set Ω. Suppose that only the identity fixes two or more points of Ω. Then the set of elements without fixed points together with the identity form a normal subgroup of order n.*

In order to prove this theorem, we reduce the hypothesis to a purely group theoretic statement. Let $\alpha \in \Omega$ and let H be the stabiliser of α. Then $[G:H] = n$. If $x \in N_G(H)$ and $h \in H - \{1\}$, then $\alpha = \alpha^{xhx^{-1}}$. So α^x is fixed by h, whence $\alpha^x = \alpha$. Thus $N_G(H) = H$. Since the intersection of two distinct point stabilisers is trivial, it follows that, if $g \in G$, then either $H^g \cap H = 1$ or $H^g \cap H = H$.

Thus we need to establish the following result, noting that if the hypotheses are satisfied, then the action of G on the right cosets of the subgroup H will satisfy the hypothesis of Theorem 33.

We shall use the following standard notation in the remainder of this book: if H is a subgroup of a group G, then the set of nonidentity elements of H will be denoted by $H^\#$.

Theorem 34. *Let G be a finite group and let H be a nonidentity subgroup of G such that $H \cap H^x = 1$ for each element $x \in G - H$. Then H has a normal complement N in G.*

Proof. If h_1, $h_2 \in H^\#$ and $h_1^g = h_2$, then it is clear that $g \in H$. Also, if X denotes the set of elements of G which are *not* conjugate to any element of $H^\#$, then we may count elements to show that $|X| = [G:H]$. So it will be sufficient to show that the set X is a subgroup. We shall do this by displaying X as the intersection of the kernels of certain characters of G.

Let $\theta, \xi \in \mathscr{C}(H, \mathbb{C})$. Then, if $g \in G$, it is readily seen from Corollary 31 that

$$\theta^G(g) = \left\{ \begin{array}{ll} 0 & \text{if } g \in X - \{1\}, \\ [G:H]\theta(1) & \text{if } g = 1, \\ \theta(g) & \text{if } g \in H^\#. \end{array} \right\} \tag{4.1}$$

If we compute the inner product of θ^G and ξ^G, we obtain the formula

$$(\theta^G, \xi^G)_G = |G|^{-1}\left(\theta^G(1)\xi^G(1) + [G:H] \sum_{g \in H^\#} \theta^G(g)\overline{\xi^G(g)} \right).$$

So, if $\xi(1) = 0$, it follows that

$$(\theta^G, \xi^G)_G = (\theta, \xi)_H. \tag{4.2}$$

Let $\varphi_1, \ldots, \varphi_t$ be the distinct irreducible characters of H, with $\varphi_1 = 1_H$. For $2 \leqslant i \leqslant t$, define generalised characters θ_i of H by

$$\theta_i = \varphi_i - \varphi_i(1) \cdot \varphi_1.$$

Since $\theta_i(1) = 0$, we see from (4.2) that

$$\| \theta_i^G \|^2 = (\theta_i^G, \theta_i^G)_G = (\theta_i, \theta_i)_H = 1 + \varphi_i(1)^2.$$

If $\chi_1 = 1_G$, then the Frobenius reciprocity theorem yields that

$$(\theta_i^G, \chi_1)_G = (\theta_i, \chi_1|_H)_H = (\theta_i, \varphi_1)_H = -\varphi_i(1).$$

So, as $\theta_i^G(1) = 0$, for each i there must be an irreducible character χ_i of G for which

$$\theta_i^G = -\varphi_i(1)\cdot\chi_1 + \chi_i.$$

Now, applying (4.1), we may compute χ_i, and obtain the values

$$\chi_i(g) = \left\{ \begin{array}{ll} \varphi_i(1) & \text{if } g \in X - \{1\}, \\ \varphi_i(1) & \text{if } g = 1, \\ \varphi_i(g) & \text{if } g \in H^\#. \end{array} \right\} \tag{4.3}$$

For each element $g \in H^\#$, there is some character φ_i of H for which $\varphi_i(g) \neq \varphi_i(1)$. Hence

$$X = \bigcap_{i=2}^{t} \ker \chi_i,$$

and this shows that X is a subgroup.

We note that the formulae (4.1) and (4.2) depended only on the fact that $H \cap H^x = 1$ if $x \notin H$. We also note that, in (4.3), we were able to show that $\chi_i(g) = \varphi_i(g)$ whenever $g \in H$. Since two elements of H are conjugate in G if and only if they are already conjugate in H, this means that part of the character table of G is identical with the corresponding part of that for H. These are the central ideas behind the theory of exceptional characters which we shall develop in detail in Chapters 3 and 4.

A group which satisfies the hypothesis and conclusion of Theorem 34 is called a *Frobenius group*. The subgroup N is called the *Frobenius kernel* of G, and H is called a *Frobenius complement*. Since every element of G lies either in N or in a unique conjugate of $H^\#$, it is an easy exercise to verify the following properties of a Frobenius group.

Proposition 35. *Let G be a group and let H and N be subgroups with N normal, $G = HN$ and $H \cap N = 1$. Then the following statements are equivalent.*

(i) *$H \cap H^x = 1$ whenever $x \notin H$.*
(ii) *If $h \in H^\#$, then $h^g \in H$ if and only if $g \in H$.*
(iii) *$C_G(n) \subseteq N$ for all $n \in N^\#$.*

In view of (iii), if $h \in H^\#$, then $C_N(h) = 1$. Hence conjugation by the element

h induces a fixed-point-free automorphism on N. In particular, we may choose h to be an element of prime order. Then Thompson's proof of the Frobenius Conjecture (Theorem 1.26) yields the following.

Corollary 36. *The kernel of a Frobenius group is nilpotent.*

The structure of a Frobenius complement is also severely restricted, and we refer to Huppert [H; Ch. 5, §8, pp. 495ff.] for details. In the following exercises we give some basic group theoretic properties of Frobenius groups. In Section 6, we shall determine some further properties of Frobenius groups, including their irreducible characters. Later, we shall be considering situations where Frobenius groups occur as subgroups of a given group, and we shall need such information to determine consequences for the characters of the whole group.

Exercises

1. Prove Proposition 35.
2. Show that the kernel and complement of a Frobenius group have coprime orders.
3. Let $G = HN$ be a Frobenius group as in Proposition 35. Show that every abelian subgroup of H is cyclic by establishing the following.

   (i) Let M be an H-invariant elementary abelian q-subgroup in $Z(N)$. Suppose that H_0 is a subgroup of H which is elementary abelian of order p^2 for some prime p. Show that M has the structure of a $\mathbb{Z}_q H_0$-module.

   (ii) Let $F = \mathbb{Z}_q(\omega)$ where ω is a pth root of unity. Show that an irreducible FH_0-module M_0 is one-dimensional.

   (iii) Show that some nonidentity element of H_0 must act trivially on M_0. Hence obtain a contradiction.

   [Remark. It follows immediately that any Sylow subgroup of H contains a unique subgroup of prime order. Since a nonabelian group of order p^3 for p odd necessarily contains noncentral elements of order p, Sylow subgroups of H of odd order are cyclic. A Sylow 2-subgroup is either cyclic or a (generalised) quaternion group.]

4. Show that the Frobenius kernel in a Frobenius group G is a uniquely determined subgroup and that any two permutation representations of G as a Frobenius group are equivalent (as permutation representations).

Is there another way of showing that two such representations are equivalent as *linear* representations?

5 Induction from normal subgroups

Let H be a normal subgroup of the group G. If χ is a character of H and $g \in G$, then we may define a character χ^g of H by

$$\chi^g(x) = \chi(g^{-1}xg).$$

This is consistent with the notation introduced in Section 1 where g acts on H as an inner automorphism. If M is a $\mathbb{C}H$-module affording χ, then we may readily verify that χ^g is afforded by the conjugate module $M^{(g^{-1})}$: indeed, we may identify $M^{(g^{-1})}$ with the summand $M \otimes g^{-1}$ of M^G. We may therefore make analogous definitions for characters as for modules, and define the *inertia subgroup* $I(\chi)$ by

$$I(\chi) = \{g \in G \mid \chi^g = \chi\}$$

and call the characters χ^g the *conjugates* of χ. Then $I(M) = I(\chi)$ and Theorem 1.39 and Corollary 1.40 immediately translate to the following.

Theorem 37. *Let H be a normal subgroup of G and suppose that χ is a character of H. Then*

$$\chi^G(x) = \begin{cases} r\sum \chi^g(x) & \text{if } x \in H, \\ 0 & \text{otherwise,} \end{cases}$$

where $r = [I(\chi):H]$ and the sum is carried over the distinct conjugates of χ, which may be taken as $\{\chi^{g_i^{-1}}\}$ where $G = \bigcup_i I(\chi)g_i$.

Corollary 38. *With the notation above, suppose also that χ is irreducible. Then χ^G is irreducible if and only if $I(\chi) = H$.*

We now turn to the relationship between the irreducible characters of H and those of G.

Theorem 39. *Let H be a normal subgroup of G and let ζ be an irreducible character of G. Then there exist an irreducible character χ of H and a positive integer e such that*

(i) $\zeta|_H = e\sum \chi^g$ *where the sum is carried over the distinct conjugates of χ, and*

(ii) *if η is an irreducible constituent of $\zeta|_{I(\chi)}$ for which χ is a constituent of $\eta|_H$, then $\zeta = \eta^G$ and $\eta|_H = e\chi$.*

(Note. The integer e is called the *ramification index* of χ in H.)

Proof. Let χ be an irreducible constituent of $\zeta|_H$. Then part (i) is an immediate consequence of Clifford's Theorem. (For an alternative proof, see Exercise 4.)

Now take η as in (ii). Since χ has no other conjugate under $I(\chi)$, part (i) shows that $\eta|_H = f\chi$ for some integer f. Suppose that $[G:I(\chi)] = s$ and $\chi(1) = t$. Then χ has s distinct G-conjugates $\chi_1, \dots, \chi_s$ and $\zeta|_H = e(\chi_1 + \cdots + \chi_s)$. Since η is a constituent of $\zeta|_{I(\chi)}$, we see that $f \leqslant e$. Now let $\tilde{\eta}$ be an irreducible constituent of $\zeta|_{I(\chi)}$ of least possible degree. Then

$$\tilde{\eta}(1) \leqslant \eta(1) = ft \leqslant et. \tag{5.1}$$

Now $\zeta(1) = est$ and hence $\tilde{\eta}^G(1) \leqslant \zeta(1)$. However, by the Frobenius reciprocity formula,

$$(\tilde{\eta}^G, \zeta)_G = (\tilde{\eta}, \zeta|_{I(\chi)})_{I(\chi)} \geqslant 1.$$

So $\tilde{\eta}^G = \zeta$ and we may therefore replace each of the inequalities in (5.1) by equalities. In particular, for our *original* choice, $\eta^G = \zeta$ and $\eta|_H = e\chi$.

We complete this section with two applications; then in the next section we shall discuss the irreducible characters of Frobenius groups. Our first application is a result about the characters of nilpotent groups which we shall require in Chapter 5 for Brauer's characterisation of characters.

Theorem 40. *Let G be a nilpotent group and suppose that ζ is an irreducible character of G. Then there are a subgroup H of G and a linear character φ of H such that $\zeta = \varphi^G$.*

Proof. We apply induction to $|G|$. If $\zeta(1) > 1$ we may suppose that $\ker \zeta = 1$: otherwise we consider characters of $G/\ker \zeta$.

Let N be an abelian normal subgroup with $N \nsubseteq Z(G)$; by Theorem 39(i),

$$\zeta|_N = e \sum \chi^g$$

where the sum is carried over the distinct conjugates of some irreducible, and hence linear, character χ of N. If χ were conjugate only to itself under G, then $\zeta|_N = \zeta(1)\chi$ and so $N \subseteq Z(G)$, which is not the case. Hence $I(\chi) \neq G$. Now let η be a character of $I(\chi)$, chosen as in Theorem 39(ii). Then $\zeta = \eta^G$. However, by induction, $\eta = \varphi^{I(\chi)}$ for some linear character φ of a subgroup H of $I(\chi)$. But then $\zeta = \varphi^G$.

We remark that a group for which every irreducible character is induced from a linear character of some subgroup is called a *monomial group* (M-group). It is known that a monomial group is necessarily soluble, but we shall not investigate this point further.

Table 2.1

	1	2	4	3	7_1	7_2
$\|C(x)\|$	168	8	4	3	7	7
χ_1	1	1	1	1	1	1
χ_2	7	-1	-1	1	0	0
χ_3	8	0	0	-1	1	1
χ_4	6	2	0	0	-1	-1
χ_5	3	-1	1	0	α	β
χ_6	3	-1	1	0	β	α

To complete this section, we shall construct the character table of the group PGL(2, 7) as an explicit example of an application of these results, starting from the character table of PSL(2, 7) whose construction was set as an exercise in Section 1. The character table of PGL(2, 7) will be used to demonstrate a recent application of character theory in Section 11.

The irreducible characters of PSL(2, 7) are given in Table 2.1. The conjugacy classes are labelled by the orders of their elements. The orders of centralisers are easily computed either from SL(2, 7) or from GL(3, 2), and we note that a Sylow 7-normaliser is a Frobenius group of order 21.

The characters χ_2 and χ_4 arise from the doubly transitive actions of PSL(2, 7) on the projective line and of GL(3, 2) on the seven nonzero vectors of a vector space, respectively. The remainder of the character table is easily determined using the orthogonality relations, where α and β are the sums $\omega + \omega^2 + \omega^4$ and $\omega^{-1} + \omega^{-2} + \omega^{-4}$ where ω is a primitive seventh root of 1.

The matrix $\begin{pmatrix} -1 & 0 \\ 0 & 1 \end{pmatrix}$ induces an outer automorphism of order 2 on PSL(2, 7) by conjugation in GL(2, 7), interchanging the two conjugacy classes of elements of order 7. Let σ be the corresponding element of PGL(2, 7). Then it is easy to see that $C(\sigma) \cong Z_2 \times D_6$, a Sylow 7-normaliser is a Frobenius group of order 42, and an element of order 6 generates its own centraliser.

The action of σ must interchange χ_5 and χ_6 and fix every other irreducible character: hence, by Corollary 38, χ_5 and χ_6 induce the same irreducible character ψ of degree 6 in PGL(2, 7). Let χ be any other nonlinear irreducible character of PSL(2, 7), inducing a character $\tilde{\chi}$ of PGL(2, 7). Then $I(\chi) = \text{PGL}(2, 7)$ and

$$\tilde{\chi}(g) = \begin{cases} 2\chi(g) & \text{if } g \in \text{PSL}(2, 7), \\ 0 & \text{otherwise;} \end{cases}$$

Table 2.2

	1	2	4	3	7	σ	y_1	y_2	6
$\lvert C(g)\rvert$	336	16	8	6	7	12	8	8	6
φ_1	1	1	1	1	1	1	1	1	1
φ_2	7	-1	-1	1	0	1	-1	-1	1
φ_3	8	0	0	-1	1	2	0	0	-1
φ_4	6	2	0	0	-1	0	$\sqrt{2}$	$-\sqrt{2}$	0
ψ	6	-2	2	0	-1	0	0	0	0
θ	1	1	1	1	1	-1	-1	-1	-1
φ_2'	7	-1	-1	1	0	-1	1	1	-1
φ_3'	8	0	0	-1	1	-2	0	0	1
φ_4'	6	2	0	0	-1	0	$-\sqrt{2}$	$\sqrt{2}$	0

hence $\|\tilde{\chi}\|^2 = 2$. If φ is an irreducible constituent of $\tilde{\chi}$, then χ is a constituent of $\varphi|_{\mathrm{PSL}}$: hence $\varphi(1) \geqslant \chi(1)$ and so $\tilde{\chi} = \varphi + \varphi'$ where $\varphi|_{\mathrm{PSL}} = \varphi'|_{\mathrm{PSL}}$ and $\varphi(1) = \varphi'(1) = \chi(1)$. Thus χ can be extended to two irreducible characters of PGL(2, 7). In fact, since PGL(2, 7) has a unique nonprincipal linear character θ, we have $\varphi' = \theta\varphi$.

We have thus produced nine distinct irreducible characters of PGL(2, 7) and, by summing the squares of degrees, we see that there can be no more. So PGL(2, 7) has four conjugacy classes lying outside PSL(2, 7), those of σ, the elements of order 6 in $C(\sigma)$, and two unknown classes for which we take representatives y_1 and y_2. Their centralisers must have orders prime to 7 and hence, by counting elements, $|C(y_1)| = |C(y_2)| = 8$. If we now attempt to determine the complete character table of PGL(2, 7) using the above information and the orthogonality relations alone, we find that all the character values may be computed except for the values of φ_4 and φ_4' on y_1 and y_2. Here the orthogonality relations will give values of either $\pm\sqrt{2}$ or $\pm i\sqrt{2}$, but will not allow us to decide which. However, both of these sets of values lie in $\mathbb{Q}(\sqrt[8]{1})$ but not $\mathbb{Q}(\sqrt[4]{1})$. Hence the elements y_1 and y_2 have order 8. But now, if y_1 and σ lie in the same Sylow 2-subgroup, $y_1\sigma$ must have order 2 and hence σ inverts y_1. So the character values are real. (A matrix calculation in GL(2, 7) would show directly that PGL(2, 7) has dihedral Sylow 2-subgroups, and a further alternative argument will be given as an application of Theorem 48 in Section 8.) The character table is therefore completely determined, and is given in Table 2.2.

Exercises

1. Prove Theorem 37 and an analogue for class functions by a direct

calculation using the formula for induced class functions. Deduce Corollary 38.

2. Let H be a normal subgroup of G and let χ be a character of H. Show that $(\chi^g)^G = \chi^G$ for any conjugate χ^g of χ.
3. Let H be a normal subgroup of G and suppose that $G = \bigcup_{i=1}^n Hg_i$. If χ_1 and χ_2 are characters of H, show that

$$(\chi_1^G, \chi_2^G)_G = \sum_{i=1}^n (\chi_1^{g_i^{-1}}, \chi_2)_H.$$

In particular, if χ is an irreducible character of H, show that

$$(\chi^G, \chi^G) = [I(\chi):H]$$

and deduce that χ is a constituent of $\chi^G|_H$.

[Note. This may be obtained either from the intertwining number theorem or, alternatively, by direct computation from Theorem 37.]

4. Deduce from Exercise 3 that, if χ and ψ are irreducible characters of H which are not conjugate under the action of G, then $(\chi^G, \psi^G) = 0$. Deduce from this the statement of Theorem 39(i).
5. Let G be a nonabelian group all of whose irreducible characters have degree a power of a fixed prime p. Prove that G has a normal subgroup H of index p. Use Exercise 3 to show that every irreducible character of H has degree a power of p and deduce that G has an abelian normal p-complement.

[This is a special case of a theorem of Thompson.]

6. Show that a metabelian group is an M-group.
7. Determine which of the characters φ_2 and φ_2' arise from the action of $\mathrm{PGL}(2,7)$ on the projective line. Why does χ_4 not admit a similar extension?

Carry out the details of the computation of the character table of $\mathrm{PGL}(2,7)$.

6 Frobenius groups

Since Frobenius groups will play a central role in our applications of character theory, we shall collect together some group theoretical facts that we shall need for the next two chapters. We shall then apply the results of the previous section in order to give a description of the irreducible characters of these groups. For the convenience of our later applications, we shall change our notation from that of Section 4.

Proposition 41. *Let G be a Frobenius group with Frobenius kernel H and complement E. Put $|E| = e$. The following properties hold.*

(i) *Let x be a fixed element of $E^{\#}$. Then $H = \{[x,h] \mid h \in H\}$.*
(ii) *If e is even, then H is abelian.*
(iii) *If e is odd, then E is metacyclic, and the Sylow subgroups of E are cyclic.*

Proof. If $h_1, h_2 \in H$ and $[x,h_1] = [x,h_2]$, then $h_1^{-1}xh_1 = h_2^{-1}xh_2$ and so $(h_2h_1^{-1})x = x(h_2h_1^{-1})$. Since $C_H(x) = 1$, this forces $h_1 = h_2$ so that the number of distinct commutators of the form $[x,h]$ is $|H|$.

To establish (ii), we must show that a finite group which admits a fixed-point-free automorphism of order 2 is abelian. This is a classical result due to Burnside, but we shall sketch a proof more in the spirit of those group theoretical methods described in this book.

Let X be a nonabelian group of minimal order subject to admitting a fixed-point-free automorphism α of order 2. Then every proper α-invariant subgroup and quotient is abelian. We first claim (without resort to Thompson's theorem) that X is nilpotent. For, if P were a nonnormal α-invariant Sylow p-subgroup of X, then $N_X(P)$ would be abelian and X would have an abelian normal p-complement N by Burnside's transfer theorem (Corollary 1.48). But then X would be soluble and thus nilpotent, formally a contradiction. In fact, we have shown that X must be a p-group of class 2 by the minimality of our choice. In any α-invariant abelian subgroup or quotient, α fixes every element of the form $g^\alpha g$: hence α acts on such sections by inversion. Now, if $x \in X - Z(X)$, it follows that $x^\alpha = x^{-1}z$ for some $z \in Z(X)$ and

$$x = x^{\alpha^2} = (x^{-1})^\alpha z^\alpha = xz^{-2}.$$

Since X must have odd order, $z = 1$. If $y \in X$, then

$$xy = (x^{-1}y^{-1})^\alpha = (x^{-1}y^{-1})^{-1} = yx:$$

hence X is abelian. Again, this is formally a contradiction, and (ii) has been established.

By Exercise 3 of Section 4 (and the remark following), every abelian subgroup of E is cyclic. Suppose that e is odd. Then every Sylow subgroup of E is cyclic. We may successively transfer out Sylow subgroups by primes in ascending order and see that E is soluble. Let N be a maximal normal abelian subgroup of E. Then N is cyclic and $N = C_E(N)$. Now N has an abelian automorphism group and, since homomorphic images of E have cyclic Sylow subgroups, E/N is cyclic.

We can now give a refinement of Proposition 35.

Proposition 42. *Let G be a group having a proper subgroup H such that $C_G(h) \subseteq H$ for all $h \in H^{\#}$. Then H is a Hall subgroup of G. If, in addition, H is normal in G, then G is a Frobenius group with Frobenius kernel H.*

Proof. Let P_0 be a Sylow p-subgroup of H and let P be a Sylow p-subgroup of G containing P_0. Then $Z(N_P(P_0)) \cap P_0 \neq 1$ and

$$N_P(P_0) \subseteq C_G(Z(N_P(P_0)) \cap P_0) \subseteq H.$$

Hence P_0 is a Sylow p-subgroup of G and H is a Hall subgroup of G.

If $H \lhd G$, then it suffices, in view of Proposition 35, to show that H has a complement E. The arguments of parts (ii) and (iii) of the previous proof apply to the quotient G/H. If G/H has even order, then $Z(G/H)$ has even order, while if G/H has odd order, G/H is metacyclic and hence has a normal subgroup of prime order. In either case, G/H has a normal subgroup L/H of prime order p where $p \nmid |H|$ and, if P is a Sylow p-subgroup of L, we may take $E = N_G(P)$ by a Frattini argument.

Next we consider the irreducible characters of a Frobenius group G. Let $G = HE$. Those characters which have H in their kernels are given by the characters of E: we shall not give a detailed description of them, although we will remark that if E is metacyclic, then they are all monomial and can therefore be described by the methods of the last section. However, it is the irreducible characters which do not contain H in their kernels with which we will be primarily concerned.

Theorem 43. *Let φ be a nonprincipal irreducible character of H. Then φ^G is an irreducible character of G which vanishes outside H, and every irreducible character of G which does not contain H in its kernel arises in this way. The number of such characters of G is equal to the number of G-conjugacy classes of nonidentity elements in H and, if χ is any such character of G, then e divides $\chi(1)$ and $\chi(1) \leqslant e[H:Z(H)]^{1/2}$. Furthermore, some such character has degree e.*

Proof. If $x \notin H$, then $C_H(x) = 1$ and, since $(|E|, |H|) = 1$, no nonidentity conjugacy class is left invariant under conjugation by x. By Theorem 12, no nonprincipal irreducible character of H is fixed by x so that $I(\varphi) = H$. By Corollary 38, φ^G is irreducible and, by the last remark, the number of distinct characters of G which can be constructed in this way is equal to the number of G-conjugacy classes of elements in H. Now Theorem 39(ii) shows that every irreducible character χ with $H \nsubseteq \ker \chi$ must arise in this way: in particular, e divides $\chi(1)$ and, if φ is a linear character of H, then $\varphi^G(1) = e$.

Finally, if $g \in Z(H)$, then $|\varphi(g)| = \varphi(1)$: hence, by the orthogonality relations, we see that

$$|H|^{-1} \sum_{g \in Z(H)} |\varphi(g)|^2 = [H:Z(H)]^{-1} \varphi(1)^2 \leqslant (\varphi, \varphi)_H = 1.$$

Thus $\varphi(1) \leqslant [H:Z(H)]^{1/2}$ and so $\varphi^G(1) \leqslant e[H:Z(H)]^{1/2}$.

Exercises

1. Let X be a nonnilpotent group of order 24 having a normal subgroup N isomorphic to Q_8, and let t be an element of order 3. Let ψ be the faithful character of N. Show that ψ^X has three constituents, each of degree 2, and that there is one, say χ, such that $\chi(t) = -1$. Show that in the representation affording χ, no nonidentity element has a fixed point.

   Let p be a prime such that $p \equiv 1 \pmod 4$. Determine an analogous representation of X over $\mathbb{Z}_p$ and deduce that X may occur as a Frobenius complement.

   Deduce that a Frobenius complement of even order need not be metacyclic, even if it is soluble.

2. Let χ be an irreducible character of $\mathrm{SL}(2,5)$ of degree 2. (See Exercise 13 of Section 1.) Show that, in a representation affording χ, no nonidentity element fixes a nonzero vector.

   Let p be a prime, $p \geqslant 7$. By regarding χ as the character of a representation over a suitable finite extension of $\mathbb{Z}_p$, show that $\mathrm{SL}(2,5)$ can occur as a Frobenius complement.

   [Note. A different proof can be given as a result of the next exercise. In Exercise 5 of Section 9, we shall show that, if $q \equiv \pm 1 \pmod 5$, then $\mathrm{PSL}(2,q)$ contains a subgroup isomorphic to $\mathrm{PSL}(2,5)$.]

3. Show that any subgroup of $\mathrm{SL}(2,q)$ of order coprime to q acts as a Frobenius complement in the natural action on an elementary abelian group of order q^2.

4. Let $G = HE$ be a Frobenius group. Show that, unless $|E| = |H| - 1$, any irreducible character of G which is constant on $H^{\#}$ must contain H in its kernel.

   [Hint. Show first that the space of class functions on H with this property has dimension 2. Then show that, if some irreducible character of G which does not contain H in its kernel has this property, then all nonprincipal characters of H must be conjugate.]

7 The special linear groups SL(2, 2ⁿ)

As an application of the results and techniques discussed in the preceding sections of this chapter, we shall determine the character tables of the groups $\mathrm{SL}(2,2^n)$ for $n \geqslant 2$. This exercise will serve two purposes. First it will provide an introduction to at least some of the more elementary ideas that are involved in determining the characters of the finite simple groups of Lie type: the deeper ideas due to Deligne and Lusztig are well beyond the scope of this book, and the interested reader may refer to the recent book by Carter [C]. But secondly, and more critically for our purposes, it will give some of the flavour of the calculations involved when character theory is combined with group theory as in the applications to the classification of finite simple groups. We shall be concerned both with the techniques of calculation and in seeing, in reverse, the way in which character theoretic information can be exploited to determine group theoretic properties. The latter will lead in this section to a characterisation of these groups by their character tables: a further characterisation, starting with less information, will be given in Section 10.

Let $G = \mathrm{SL}(2, q)$ with $q = 2^n$, $n \geqslant 2$. It is easy to show that G is simple although we shall not need this fact; actually, we can deduce it once we see that every nonprincipal irreducible character is faithful. We shall only sketch the steps to determine the character table of G: verification of the details will be left as an exercise.

Let U be the subgroup of lower unitriangular matrices and let H be the subgroup of diagonal matrices in G. Let $\sigma = \begin{pmatrix} 0 & 1 \\ 1 & 0 \end{pmatrix}$. Then the following statements are easily checked.

(I) $|G| = q(q-1)(q+1)$, $|U| = q$ and $|H| = q-1$.

(II) Every element of $U^\#$ is an involution, and all such involutions are conjugate under the action of H.

(III) If $\tau \in U^\#$, then $C_G(\tau) = U$ and $N_G(U) = HU$. In fact, $N_G(U)$ is a Frobenius group with Frobenius kernel U.

(IV) H is cyclic. If $x, x' \in H^\#$, then $C_G(x) = H$ and, if x, x' are conjugate in G, then $x' = x^{\pm 1}$: furthermore, $N_G(H) = \langle H, \sigma \rangle$.

Since $q-1$, q and $q+1$ are mutually coprime, any element of order dividing $q(q-1)$ is conjugate in G to an element of either H or U: consequently, any other element has order dividing $q+1$ and centraliser of order also dividing $q+1$. In fact, we claim

(V) G contains a cyclic subgroup K of order $q+1$ such that, if $y, y' \in K^\#$ and y, y' are conjugate in G, then $y' = y^{\pm 1}$: furthermore, $C_G(y) = K$ and $N_G(K)$ is dihedral of order $2(q+1)$. Any element of G of order dividing $q+1$ is conjugate to an element of K.

To establish this, we first embed G into $G^* = \mathrm{SL}(2, q^2)$. G^* has order $q^2(q^2-1)(q^2+1)$ and any element of G^* of order dividing $q+1$ must be conjugate to an element of the diagonal subgroup H^* and we may apply (IV): then any nonidentity element of order dividing $q+1$ in G has cyclic centraliser, and the only possible distinct power to which it can be conjugate is its inverse. It will therefore be sufficient to display a dihedral subgroup in G of order $2(q+1)$.

Table 2.3

	1	τ	$x \in H^{\#}$	$y \in K^{\#}$
1	1	1	1	1
χ	q	0	1	-1
$\Phi_i,\ i=1,\ldots,\frac{1}{2}(q-2)$	$q+1$	1	$\varphi_i(x)$	0
$\Psi_j,\ j=1,\ldots,\frac{1}{2}q$	$q-1$	-1	0	$-\psi_j(y)$

Write $\mathrm{GF}(q) = A \cup A^*$ where

$$A = \{\lambda \mid \lambda = \alpha^2 + \alpha \text{ for some } \alpha \in \mathrm{GF}(q)\}$$

and A^* is its complement. Then $|A| = |A^*| = \frac{1}{2}q$. For $\mu \in A^*$, the polynomial $t^2 + t + \mu$ is irreducible over $\mathrm{GF}(q)$, and hence the companion matrix

$$\begin{pmatrix} 0 & 1 \\ 1 & \mu \end{pmatrix} \tag{7.1}$$

has order dividing $q+1$. By consideration of degrees we see that

$$t^{q+1} - 1 = (t-1) \prod_{\mu \in A^*} (t^2 + t + \mu):$$

since G^* contains an element of order $q+1$ having a matrix of the shape (7.1) as its rational canonical form, *some* matrix of the form (7.1) has order $q+1$ and, as

$$\sigma^{-1}\begin{pmatrix} 0 & 1 \\ 1 & \mu \end{pmatrix}\sigma = \begin{pmatrix} \mu & 1 \\ 1 & 0 \end{pmatrix} = \begin{pmatrix} 0 & 1 \\ 1 & \mu \end{pmatrix}^{-1},$$

we have found the desired dihedral group.

In particular, we have determined the conjugacy classes of G: they are

(VI) (i) 1,
(ii) one conjugacy class of involutions,
(iii) $\frac{1}{2}(q-2)$ conjugacy classes of elements of orders dividing $q-1$,
(iv) $\frac{1}{2}q$ conjugacy classes of elements of orders dividing $q+1$.

Let the nonlinear characters (of degree 2) of $N_G(H)$ be $\{\varphi_i\}$ and of $N_G(K)$ be $\{\psi_j\}$. Then the character table of G is as shown in Table 2.3. Again, we shall just sketch the steps required to show this.

G acts doubly transitively on the $q+1$ points of the projective line: if the permutation character is π, then we may compute χ from $\pi = 1 + \chi$.

Let $B = HU$. Then B is the stabiliser of a point of the projective line, and G has the double-coset decomposition $G = B \cup B\sigma B$ by double transitivity. Let $\xi_1, \ldots, \xi_{q-2}$ be the nonprincipal linear characters of B. Since $B^\sigma \cap B = H$, a direct application of Mackey's decomposition theorem (Exercise 15 of Section 3) will show that each induced character ξ_i^G is irreducible; by suitable labelling, these give the characters Φ_i and their values.

Now there are $\frac{1}{2}q$ irreducible characters $\{\Psi_j\}$ yet to be determined. Since, for $x \in H^\#$,

$$1 + 1 + \sum_i |\varphi_i(x)|^2 = q - 1 = |C_G(x)|,$$

every Ψ_j vanishes on $H^\#$: as $(\Psi_j|_H, 1_H) \in \mathbb{Z}$, it follows that $q-1$ divides $\Psi_j(1)$ for each j. Summing the squares of the character degrees, we obtain the equation

$$1 + q^2 + \tfrac{1}{2}(q-2)(q+1)^2 + \sum_j \Psi_j(1)^2 = q(q-1)(q+1),$$

from which we deduce that

$$\Psi_j(1) = q - 1$$

for all j. In particular, $\Psi_j(1)$ is odd so that $\Psi_j(\tau) \neq 0$: applying the orthogonality relations to the first two columns of the character table, we see that

$$\Psi_j(\tau) = -1$$

for all j.

It remains to determine the values of Ψ_j on $K^\#$. Let ψ be any ψ_j. Then

$$\psi^G(g) = \begin{cases} q(q-1) & \text{if } g = 1, \\ \psi(g) & \text{if } g \in K^\#, \\ 0 & \text{if } g \in B^\#. \end{cases} \tag{7.2}$$

From these values, we can compute the norm

$$\|\psi^G\|^2 = q - 1$$

while Frobenius reciprocity gives the inner products

$$(\psi^G, 1_G) = 0,$$

$$(\psi^G, \chi) = 1$$

and, for each i,

$$(\psi^G, \Phi_i) = 1.$$

Let $a_j = (\psi^G, \Psi_j)$. Then the above multiplicities imply that

$$\sum_j a_j^2 = \tfrac{1}{2}(q-2)$$

while, by consideration of degrees,

$$\sum_j a_j = \tfrac{1}{2}(q-2).$$

Thus $a_j = 1$ for all but one value of j, for which $a_j = 0$. For given j, since the induced characters ψ_j^G are all different, we may label $\{\Psi_j\}$ so that

$a_j = 0$ for each j. Then

$$\psi_j^G = \chi + \sum_i \Phi_i + \sum_k \Psi_k - \Psi_j. \tag{7.3}$$

From the orthogonality relations, we see that, if $y \in K^{\#}$,

$$\sum_k \Psi_k(y) = 1:$$

thus, from the computed values (7.2) of ψ_j^G, we finally obtain the values

$$\Psi_j(y) = -\psi_j(y)$$

for $y \in K^{\#}$.

We remark, slightly obliquely, that the significance of the final stages of this calculation, and in particular the calculation of the a_js and the choices of the Ψ_js, will become clearer in Chapters 3 and 4; in particular, the explicit formula (7.3) should be compared with the formula in the conclusion of Theorem 3.11.

Next we may ask what the character table tells us in turn about the group $\mathrm{SL}(2, 2^n)$. The answer is everything! Namely, it is possible to show the following.

Theorem 44. *Let X be a group having the same character table as* $\mathrm{SL}(2, 2^n)$ *for some $n \geqslant 2$. Then $X \cong$* $\mathrm{SL}(2, 2^n)$.

Notice that there can be no universal theorem of this type: in the exercises of Section 1 we saw that the groups D_8 and Q_8 have the same character tables. However, it *is* the case that no two simple groups have the same character tables, and direct characterisations have been given in a number of cases.

Most of the proof of Theorem 44 is beyond the scope of this book, but we shall indicate the major conclusion that can be drawn from the character table. Again, put $q = 2^n$. First, $|X|$ is known and then, by computing the orders of centralisers of elements, we deduce that X has only one conjugacy class of elements of even order (having centralisers of order q) and hence that X has an elementary abelian Sylow 2-subgroup S. By Burnside's lemma (Lemma 1.47), it follows that all involutions in S are conjugate in $N_X(S)$. Hence $N_X(S)$ is a Frobenius group of order $q(q-1)$ having index $q+1$ in X. Now X permutes the cosets of $N_X(S)$ transitively; the only possibility for a permutation character of degree $q+1$ is

$$\pi = 1 + \chi$$

from which it follows that the action is doubly transitive (Exercise 14 of Section 3). Furthermore, no nonidentity element of X fixes more than two points so that the 2-point stabiliser acts regularly on the remaining $q-1$ points and X acts as a sharply triply transitive permutation group.

At this point, we appeal to a classical theorem of Zassenhaus to complete the identification. (See, for example, Huppert and Blackburn [HB III; Ch. XI, §2, pp. 173ff.].) The goal is to explicitly construct a projective line over GF(q) on which X can act.

Proposition 45. *Let X be a sharply triply transitive permutation group. Then, for some q, X contains* PSL$(2,q)$ *as a subgroup of index* 1 *or* 2 *and X acts on the projective line.*

We used relatively little from the character table in order to get to a position where Zassenhaus' theorem could be invoked, and we shall see in Section 10 a further characterisation of these groups.

Exercise

1. The projective special linear groups PSL$(2,q)$ for q odd, $q \geqslant 5$, are simple of order $\frac{1}{2}q(q-1)(q+1)$. Let $q = p^n > 5$. Then

   (i) all involutions are conjugate;
   (ii) a Sylow p-subgroup S is elementary abelian and is the centraliser of each of its nonidentity elements, and $N(S)$ is a Frobenius group of order $\frac{1}{2}q(q-1)$;
   (iii) there are maximal cyclic subgroups H and K of orders $\frac{1}{2}(q-1)$ and $\frac{1}{2}(q+1)$ having dihedral normalisers;
   (iv) if $x, x' \in H^{\#}$, then $C(x) \subseteq N(H)$ and $x' = x^{\pm 1}$, and similarly for elements of $K^{\#}$.

   Following the techniques of this section, determine the character tables of the groups PSL$(2,q)$ for $q > 5$. Attempt the case $q \equiv 3 \pmod 4$ first, for then $\frac{1}{2}(q-1)$ is odd: the character table is a little different for $q \equiv 1 \pmod 4$.

8 The Frobenius–Schur indicator

Towards the end of Section 1, we raised the question as to whether or not it was possible to decide if a given representation with a real-valued character could be realised over the real field. In this section, we shall give a positive answer though we will first prove only a more limited assertion which seems sufficient for many applications.

First we divide the irreducible complex representations of a group G into three types. Let ρ be an irreducible representation with character χ. Then ρ is said to be of the *first kind* if ρ is realisable over $\mathbb{R}$, of the *second kind* if ρ is not realisable over $\mathbb{R}$ but χ is real, and of the *third kind* if χ is not real. Now define the *Frobenius–Schur indicator*

$$\nu(\chi) = |G|^{-1} \sum_{g \in G} \chi(g^2).$$

Theorem 46. $\nu(\chi) = 1, -1$ *or* 0 *according as* ρ *is of the first, second or third kind, respectively.*

The proof of this theorem will require the observation that any complex representation of a group may be realised by unitary matrices and thus, for representations of the first kind, by orthogonal matrices. (See Lemma 50.) Unless one is interested in the specific question of the realisability of real characters, it does not seem particularly useful to distinguish between representations of the first and second kinds: the indicator will tell us this from the character values rather than allow us to deduce character theoretic information from the indicator since one does not normally know enough about the representations themselves. However, it is useful to be able to distinguish representations of the third kind without a full knowledge of the character table: for example, this would have resolved the indeterminacy that occurred in the construction of the character table of PGL(2, 7) in Section 5 as we shall show later. We shall therefore first prove the following special case, leaving the separation of the first and second kinds to the end of this section.

Theorem 47. $\nu(\chi) = 0$ *if and only if* χ *is not real-valued. Otherwise,* $\nu(\chi) = \pm 1$.

Proof. Let M be a $\mathbb{C}G$-module affording the character χ and let M_S and M_A be the symmetric and antisymmetric subspaces of $M \otimes M$ respectively. Then $M \otimes M = M_S \oplus M_A$, and M_S and M_A afford characters χ_S and χ_A. (See Exercise 23 of Section 1.2.)

Let $\{m_1, \ldots, m_n\}$ be a basis for M and let $A = (a_{ij})$ be the matrix representing a given fixed element g with respect to this basis. Then, putting $m_{ij} = m_i \otimes m_j - m_j \otimes m_i$, we see that

$$m_{ij}g = \sum_{k<l} (a_{ik}a_{jl} - a_{jk}a_{il})m_{kl}$$

and hence, since $\{m_{ij} | i < j\}$ is a basis for M_A, that

$$\chi_A(g) = \sum_{i<j} (a_{ii}a_{jj} - a_{ji}a_{ij}).$$

Thus

$$2\chi_A(g) = \left(\sum_i a_{ii}\right)^2 - \sum_{i,j} a_{ij}a_{ji}$$

and so

$$2\chi_A(g) = (\operatorname{tr} A)^2 - \operatorname{tr}(A^2) = \chi(g)^2 - \chi(g^2). \tag{8.1}$$

From this it follows that

$$\nu(\chi) = (1_G, \chi^2 - 2\chi_A) = (1_G, \chi^2) - 2(1_G, \chi_A).$$

Since M_A is a summand of $M \otimes M$, we must have $(1_G, \chi_A) \leqslant (1_G, \chi^2)$. But then, if χ is not real-valued, $(1_G, \chi^2) = (\bar{\chi}, \chi) = 0$ and so $\nu(\chi) = 0$ while, if χ is real-valued, $(1_G, \chi^2) = (\chi, \chi) = 1$ and thus $\nu(\chi) = \pm 1$.

Notice that (8.1) may be rewritten as

$$\chi(g^2) = \chi_S(g) - \chi_A(g),$$

and hence

$$\nu(\chi) = (1_G, \chi_S) - (1_G, \chi_A).$$

Thus, for real-valued characters, $\nu(\chi) = 1$ or -1 according to whether or not it is M_S which contains the trivial constituent. This will provide the tool for distinguishing between representations of the first and second kinds.

Before we carry this out, we shall examine the following consequence of Theorem 47. The formula we obtain is the basis of many applications. We state it with the usual notation of the previous sections.

Theorem 48. *Suppose that the group G has t involutions. Then*

$$1 + t = \sum_{i=1}^{r} \nu(\chi_i)\chi_i(1).$$

In particular,

$$1 + t \leqslant \sum_{i=1}^{r} \chi_i(1).$$

Proof. We define a class function on G by

$$\theta(g) = |\{x \in G \mid x^2 = g\}|.$$

Then

$$\theta(g)\chi_i(g) = \sum \chi_i(h^2)$$

where the summation is carried over the set $X_g = \{h \in G \mid h^2 = g\}$. Hence

$$(\chi_i, \theta) = |G|^{-1} \sum_{g \in G} \left(\sum_{h \in X_g} \chi_i(h^2) \right) = \nu(\chi_i).$$

Now we see that

$$1 + t = \theta(1) = \sum_{i=1}^{r} \nu(\chi_i)\chi_i(1)$$

and the second statement follows from Theorem 47.

As an illustration of the way in which this result may be applied, we return to the problem that arose in the construction of the character table of PGL(2, 7) in Section 5. At the final stage there was a question as to whether the two characters of degree 8 took real or complex values on the unknown classes with representatives y_1 and y_2. The character degrees were already known – they were $1,7,8,6,6,1,7,8,6$. There are 21 involutions inside PSL(2, 7) and the involution σ has 28 conjugates. Thus PGL(2, 7) has at least 49 involutions. However, the sum of the character degrees is 50. Hence, by Theorem 48, we infer that $\nu(\chi) = 1$ for every irreducible character χ of PGL(2, 7) and also that there are no further involutions. In particular, every character is real-valued; Theorem 47 is sufficient to assure this, although the full power of Theorem 46 would actually allow us to deduce further that every irreducible representation is realisable over $\mathbb{R}$.

We now return to the distinction between representations of the first and second kinds. From the remark following the proof of Theorem 47, we see that, if ρ is an irreducible complex representation affording a real character χ and afforded by a (complex) module M, then $\nu(\chi) = 1$ if and only if M_S contains a trivial submodule. Now, in the notation of Theorem 47 and its proof,

$$M \otimes M = M_S \oplus M_A$$

and hence

$$(M \otimes M)^* \cong (M_S)^* \oplus (M_A)^*.$$

Now $(M \otimes M)^*$ can be identified with the space of bilinear maps from $M \times M$ to $\mathbb{C}$ and G acts naturally on this space via

$$(m, m')f^g = (mg, m'g)f.$$

Since all modules are completely reducible in characteristic 0, M_S contains a trivial submodule if and only if $(M_S)^*$ does. Thus we may characterise the desired distinction as follows.

Proposition 49. $\nu(\chi) = 1$ *if and only if* M *admits a* G*-invariant symmetric bilinear form.*

We must therefore show that this condition characterises irreducible representations of the first kind. First we need the result indicated at the beginning of this section.

Lemma 50. *Let* ρ *be a complex (resp. real) matrix representation of a group* G. *Let*

$$S = \sum_{g \in G} \rho(g)\overline{\rho(g)}^{\mathrm{T}},$$

where $\overline{\rho(g)}^{\mathrm{T}}$ denotes the complex conjugate transpose of $\rho(g)$. Then S is a positive definite Hermitian (resp. real, symmetric) matrix and

$$\rho(x)S\overline{\rho(x)}^{\mathrm{T}} = S \tag{8.2}$$

for each $x \in G$. In particular, ρ is similar to a representation in which every element of G is represented by a unitary (resp. orthogonal) matrix.

Proof. We consider only the complex case: the corresponding statements for real representations follow by specialisation. Clearly $\bar{S}^{\mathrm{T}} = S$ while, for any row vector v in the underlying space,

$$vS\bar{v}^{\mathrm{T}} = \sum_{g \in G} (vg)\overline{(vg)}^{\mathrm{T}} > 0$$

so that S is positive definite. Also, for $x \in G$,

$$\rho(x)S\overline{\rho(x)}^{\mathrm{T}} = \sum_{g \in G} \rho(xg)\overline{(\rho(xg))}^{\mathrm{T}} = S.$$

Since S is Hermitian and hence self-adjoint, by the spectral theorem S is similar under conjugation by a unitary matrix to a diagonal matrix with real positive diagonal entries so that, in (8.2), we may suppose this already to be the case. Writing $S = T^2$ where T is real, (8.2) becomes

$$(T^{-1}\rho(x)T)\overline{(T^{-1}\rho(x)T)}^{\mathrm{T}} = I,$$

so that ρ is similar to a unitary representation.

Thus, if ρ is of the first kind, there is a G-invariant symmetric bilinear form.

We shall only sketch a proof of the converse, leaving the details to the reader. Suppose that ρ is an irreducible representation afforded by a $\mathbb{C}G$-module M for which such a form $B(m, m')$ exists. Since M is irreducible, the form B must be nondegenerate. Now, if $(\cdot,\cdot)$ is a positive definite Hermitian form on M, for example the standard inner product with respect to a fixed basis, then we may define a G-invariant positive definite Hermitian form on M by

$$H(m, m') = \sum_{g \in G} (mg, m'g)$$

and a G-invariant bijective map $\varphi: M \to M$ by

$$B(m, m') = \overline{H(m\varphi, m')};$$

φ is 'conjugate linear' in the sense that $(cm)\varphi = \bar{c}(m\varphi)$ for $c \in \mathbb{C}$. However, φ^2 is linear, self-adjoint and positive definite so that, as in the proof of Lemma 50, the spectral theorem shows that there is a positive definite, self-adjoint, linear transformation ψ on M such that $\psi^2 = \varphi^2$: furthermore, ψ can be expressed as a polynomial in φ^2 with real coefficients.

Let $\sigma = \varphi\psi^{-1}$. Then σ is conjugate linear and $\sigma^2 = 1$. However, σ is linear when M is viewed as a vector space over $\mathbb{R}$, and, if

$$M_1 = \{m \in M \mid m\sigma = m\}$$

and

$$M_2 = \{m \in M \mid m\sigma = -m\},$$

then $M = M_1 \oplus M_2$ and $M_2 = iM_1$. Now M_1 is G-invariant and, since $M = \mathbb{C} \otimes_{\mathbb{R}} M_1$, M_1 affords a real representation equivalent to ρ; thus ρ is of the first kind.

This completes the proof of Theorem 46.

Exercises

1. Show that Maschke's theorem is a consequence of Lemma 50 for complex representations.
2. Fill in the details of the proof of Theorem 46 as follows.
   (i) Given a G-invariant bilinear form B on a $\mathbb{C}G$-module M, show that $\{m \in M \mid B(m, m') = 0 \text{ for all } m' \in M\}$ forms a submodule.
   (ii) Prove that the form H is G-invariant, positive definite and Hermitian.
   (iii) Establish the existence and properties of the map φ, either formally or by use of matrices to represent B and H.
   (iv) Prove that σ is G-invariant.
3. Determine whether the 2-dimensional irreducible representations of D_8 and Q_8 are realisable over $\mathbb{R}$ or not.
4. Show that if g is an element of odd order in a group G, then $\sum_i \nu(\chi_i)\chi_i(g) = 1$.
5. Show that, if $g \in G$, then $|\{x \in G \mid x^2 = g\}| = \sum_i \nu(\chi_i)\chi_i(g)$.
6. Use Theorem 48 to show that every irreducible representation of the group $\mathrm{SL}(2, 2^n)$ is realisable over $\mathbb{R}$. Verify that this is consistent with the result of Exercise 4, but that once the character degrees are known for this group, no new relations (beyond those coming from the orthogonality relations) arise.

9 Some counting methods

We recall from Section 1 (with the standard notation of that section) that the class algebra constants a_{ijk} were defined as nonnegative integers by the products

$$C_i C_j = \sum_k a_{ijk} C_k \tag{9.1}$$

and that a character formula for them was found in Corollary 9. We now interpret these class algebra constants group theoretically and establish methods which allow us to apply character theory in certain situations.

For a *fixed* element z of $\mathscr{C}_k$, let

$$\mathscr{A}(\mathscr{C}_i, \mathscr{C}_j; z) = \{(x, y) | x \in \mathscr{C}_i, y \in \mathscr{C}_j, xy = z\}. \tag{9.2}$$

Then we have as an immediate consequence of the way in which the product (9.1) is multiplied out,

Proposition 51. $a_{ijk} = |\mathscr{A}(\mathscr{C}_i, \mathscr{C}_j; z)|$: *in particular, the right-hand side is independent of the choice of the element z in the conjugacy class $\mathscr{C}_k$.*

This result can be exploited in situations where the set $\mathscr{A}(\mathscr{C}_i, \mathscr{C}_j; z)$ can be identified inside the group. The most important case is where the conjugacy classes $\mathscr{C}_i$ and $\mathscr{C}_j$ consist of involutions; here the following result was first exploited by Brauer and Fowler (1955), and it has played a central role ever since. (See footnote on page 99.)

Proposition 52. *Let x, y be involutions in a group G and let $H = \langle x, y \rangle$. Then H is dihedral and, putting $z = xy$,*

(i) $x^{-1}zx = y^{-1}zy = z^{-1}$, *and*
(ii) *x is conjugate to y in H if and only if z has odd order.*

Proof. The calculation of (i) is trivial, and it follows that H is dihedral since $H = \langle x, z \rangle$. If z has odd order, then H has twice odd order and all involutions in H are conjugate: if z has even order $2n$, then x has n conjugates $x, z^2x, z^4x, \ldots$ in H, none of which is y.

For a group G and an element $g \in G$, we define the *generalised centraliser*

$$C^*(g) = \{x \in G | x^{-1}gx = g^{\pm 1}\}.$$

Then, if x, y are involutions with $xy = g$, we have shown that $x, y \in C^*(g)$. Hence, if $\mathscr{C}_i$ and $\mathscr{C}_j$ are conjugacy classes of involutions,

$$\mathscr{A}(\mathscr{C}_i, \mathscr{C}_j; g) = \mathscr{A}_{C^*(g)}(\mathscr{C}_i, \mathscr{C}_j; g)$$

where the suffix indicates the restriction to $C^*(g)$ in (9.2). So, if $g \in \mathscr{C}_k$, we can compute the class algebra constant a_{ijk} from a knowledge of $C^*(g)$ together with a knowledge of which involutions of $C^*(g)$ lie in $\mathscr{C}_i$ or $\mathscr{C}_j$. In fact, if g has order at least 3, it is enough to know this for $C^*(g) - C(g)$ and, conversely, it is sometimes possible to obtain further information about involutions in a group from the structure of $C^*(g)$.

As brief illustrations of how these ideas are applied, we shall give two examples. The first will be generalised in the next section, while the second will be generalised in Chapter 3. (See Section 3.4, in particular Corollary 3.18, and the discussion at the end on this section.)

Example 1. Let G be a simple group having a Sylow 2-subgroup T which is elementary abelian of order 4. Suppose that $T = C_G(\tau)$ for all $\tau \in T^{\#}$. Then G is isomorphic to A_5.

Since G is simple, we may apply Burnside's transfer theorem (Corollary 1.48; but see also Exercise 4 of Section 1.8) to show that all involutions in G lie in a single conjugacy class $\mathscr{C}$. Let $\tau \in T^{\#}$. Since $\sum_i \chi_i(\tau)^2 = 4$, there are precisely four irreducible characters which do not vanish on τ, and the orthogonality relations force values of these on 1 and τ of the form

	1	τ
1_G	1	1
χ_2	f_2	1
χ_3	f_3	-1
χ_4	f_4	δ

where $\delta = \pm 1$ and $1 + f_2 - f_3 + \delta f_4 = 0$. Now $(\chi_i|_T, 1_T) \in \mathbb{Z}$. So we get congruences

$$f_2 \equiv 1 \quad f_3 \equiv -1 \quad f_4 \equiv \delta \tag{9.3}$$

modulo 4. Since τ can be a product of two involutions only if both lie in T also, $|\mathscr{A}(\mathscr{C}, \mathscr{C}; \tau)| = 2$. Hence the formula for the corresponding class algebra constant yields the equation

$$\frac{|G|}{16}\left(1 + \frac{1}{f_2} - \frac{1}{f_3} + \frac{\delta}{f_4}\right) = 2.$$

On the other hand, since G has no nonprincipal linear characters, $f_i > 1$ for $i = 2, 3, 4$, and the congruences (9.3) then imply that

$$1 + \frac{1}{f_2} - \frac{1}{f_3} + \frac{\delta}{f_4} > \frac{1}{3},$$

whence $|G| < 96$. But then $1 + f_2^2 + f_3^2 + f_4^2 < 96$ and so $f_2 = 5$. Since f_i divides $|G|$ for each i, this inequality also forces $f_3 = f_4 = 3$; hence $|G| = 60$. Now $T \cap T^g = 1$ if $g \notin N_G(T)$, for otherwise $\langle T, T^g \rangle \subseteq C(T \cap T^g)$; so G has five Sylow 2-subgroups. It now follows easily that $G \cong A_5$.

Example 2. Let G be a simple group containing an element g of order 3 for which $C_G(g) = \langle g \rangle$. Then G has even order and all involutions in G are conjugate.

Certainly $\langle g \rangle$ is a Sylow 3-subgroup of G and, by Burnside's transfer theorem, g must be conjugate to its inverse so that $C^*(g) \cong D_6$. In particular, G has even order. Let τ be an involution in G. Then the orthogonality relations imply that there are precisely three irreducible characters of G which do not vanish on g, and their values on 1, τ and g have the form

	1	τ	g
1_G	1	1	1
χ_2	x	u	1
χ_3	$x+1$	$u+1$	-1

for some integers x and u.

We claim that τ must be conjugate to an element of $C^*(g)$, and then the proof will be complete. If this is not the case, then no two conjugates of τ can have product g, and the character formula for the corresponding class algebra constant will yield the equation

$$\frac{|G|}{|C(\tau)|}\left(1+\frac{u^2}{x}-\frac{(u+1)^2}{x+1}\right)=0.$$

However,

$$1+\frac{u^2}{x}-\frac{(u+1)^2}{x+1}=\frac{(x-u)^2}{x(x+1)}, \tag{9.4}$$

whence $x=u$ and τ lies in the kernel of the character χ_2. But this is impossible since G is simple.

Especial attention should be given to the final part of this argument. The factorisation (9.4) and the conclusion that we draw from it when its value is zero will play an important role in a number of future calculations of this type.

We defined real elements in Section 1. Here we have used a stronger concept in that we have been interested in nonidentity elements which are not merely conjugate to their inverses, but which are actually inverted by an involution: such elements are said to be *strongly real.* Notice that if an element g is not strongly real, then for conjugacy classes $\mathscr{C}$ and $\mathscr{C}'$ of involutions, not necessarily different,

$$\mathscr{A}(\mathscr{C},\mathscr{C}';g)=\varnothing.$$

In this case, if $x \in \mathscr{C}$ and $y \in \mathscr{C}'$, we get a character formula

$$\sum_{i=1}^{r} \frac{\chi_i(x)\chi_i(y)\overline{\chi_i(g)}}{\chi_i(1)} = 0;$$

that is, we have a *relation* between character values. This relation can also be described as follows. For elements x, y of G, let X_{xy} be the column whose ith entry is $\chi_i(x)\chi_i(y)/\chi_i(1)$ and let Y_g be the column with entries $\chi_i(g)$. Then X_{xy} is orthogonal to Y_g whenever x and y are involutions and g is not strongly real. Notice that X_{xy} will also be orthogonal to $\bar{Y}_g$ in this case.

Next we shall consider counting arguments which do not necessarily involve involutions. We start with a weak analogue of Proposition 52 for elements of order 3.

Proposition 53. *Let g, h be elements of order* 3 *in a group G, and suppose that their product gh has order* 3. *Then*

(i) $\langle g, h\rangle$ *has a normal abelian subgroup of index* 3, *and*
(ii) (G. Higman) *if g and h are not conjugate in G, then there exists an element x of order* 3 *with $g, h \in C_G(x)$.*

Proof. We shall give only a sketch and leave the detailed verification to the reader.

For the first part, it is sufficient to establish the desired conclusion for the group given by the presentation

$$\langle g, h \mid g^3 = h^3 = (gh)^3 = 1\rangle.$$

It is easily checked that the subgroup

$$N = \langle [g, h], [g^{-1}, h^{-1}], g^{-1}h, gh^{-1}\rangle$$

is an abelian normal subgroup: this uses crucially the consequence

$$hgh = g^{-1}h^{-1}g^{-1}$$

of the relation $(gh)^3 = 1$, and also that $(hg)^3 = 1$. Then N has index 3 since N is precisely the kernel of the homomorphism from $\langle g, h\rangle$ onto a group of order 3 in which both g and h are sent to the same generator.

Now suppose that A is the normal abelian subgroup of index 3 in $\langle g, h\rangle$ inside G and that g and h are not conjugate. Then $|A|$ is divisible by 3, for otherwise $g \equiv h^{-1} \pmod{A}$ and hence $gh \in A$, which is impossible. Now let P be a Sylow 3-subgroup of $\langle g, h\rangle$. Then $\langle g, h\rangle = PA$ and we may choose $x \in Z(P) \cap A \subseteq C_G(\langle g, h\rangle)$.

Part (ii) of the above result is especially useful if a Sylow 3-subgroup P

of G satisfies the conditions

$$\left.\begin{array}{ll} P\cap P^g=1 & \text{if } g\notin N(P), \\ C_G(x)\subseteq N(P) & \text{whenever } x\in P^{\#}. \end{array}\right\} \qquad (\#)$$

We shall illustrate this in the next example. As with Example 2, the ideas will form a model for the general treatment which we shall discuss in the next chapter.

Example 3. Let G be a finite simple group in which the normaliser of a Sylow 3-subgroup P is a Frobenius group of order 36, and assume that the conditions (#) hold. Then $G\cong A_6$.

The Sylow 3-subgroup P is necessarily elementary abelian, and an involution in $N(P)$ inverts every element of P; hence elements of order 3 are strongly real and character values on them are real. By Burnside's lemma (Lemma 1.47), two elements of P are conjugate in G if and only if they are already conjugate in $N(P)$; hence G contains exactly two conjugacy classes of elements of order 3. Let x_1 and x_2 be representatives. As in the previous examples, the first task is to determine the nonvanishing irreducible character values on these elements; by considering $\sum_i \chi_i(x)^2$, we see that such values are ± 1 or ± 2.

Suppose that χ is an irreducible character with $\chi(x_1)=0$. Since $(\chi|_{\langle x_1\rangle}, 1_{\langle x_1\rangle})\in\mathbb{Z}$ and x_1 is conjugate to its inverse, $\chi(1)\equiv 0 \pmod 3$. Then $\chi(x_2)=0$ also from a similar consideration of $\chi_{\langle x_2\rangle}$. Thus if χ is an irreducible character which does not vanish on x_1, then χ does not vanish on x_2 either. The orthogonality relations now force the values shown in Table 2.4 for such characters where the $\delta_i=\pm 1$, and

$$1+\delta_2 u+\delta_3 v+\delta_4 w=\delta_5 z.$$

All other irreducible characters vanish on both x_1 and x_2. (Compare this argument with that indicated in Exercise 1 of Section 3.3.)

If $\mathscr{C}_1$ and $\mathscr{C}_2$ are the conjugacy classes containing x_1 and x_2, then $\mathscr{A}(\mathscr{C}_1,\mathscr{C}_2;x_1)=\mathscr{A}_P(\mathscr{C}_1,\mathscr{C}_2;x_1)$ by Proposition 53(ii). Hence

$$|\mathscr{A}(\mathscr{C}_1,\mathscr{C}_2;x_1)|=2.$$

Computing the corresponding class algebra constant from the character values shown in Table 2.4 yields the equation

$$\frac{|G|}{81}\left(1+\frac{\delta_2}{u}+\frac{\delta_3}{v}+\frac{\delta_4}{w}+\frac{2\delta_5}{z}\right)=2. \qquad (9.5)$$

Restrictions to P as in the previous examples give the congruences

$$u\equiv\delta_2,\quad v\equiv\delta_3,\quad w\equiv\delta_4,\quad z\equiv 4\delta_5 \pmod 9: \qquad (9.6)$$

Table 2.4

	1	x_1	x_2
1_G	1	1	1
χ_2	u	δ_2	δ_2
χ_3	v	δ_3	δ_3
χ_4	w	δ_4	δ_4
χ_5	z	$-2\delta_5$	δ_5
χ_6	z	δ_5	$-2\delta_5$

hence, since no nonprincipal irreducible character is linear,

$$1+\frac{\delta_2}{u}+\frac{\delta_3}{v}+\frac{\delta_4}{w}+\frac{2\delta_5}{z}\geqslant 1-\frac{1}{8}-\frac{1}{8}-\frac{1}{8}-\frac{2}{5}=\frac{9}{40}$$

and so

$$|G|\leqslant 720.$$

Now $|G|=720$ only if $u=v=w=8$ and $z=5$. But then

$$\sum_{i>6}\chi_i(1)^2=477$$

which is not divisible by 81, a contradiction since $\chi_i(1)\equiv 0 \pmod 9$ for $i>6$. Using the fact that the character degrees divide the group order, an examination of the possible solutions to the congruences (9.6) together with the equations (9.5) now forces (up to a choice of labelling) the degrees

$$u=v=8,\quad w=10,\quad z=5.$$

In particular, these degrees imply that $|G|=360$ and that there is exactly one further irreducible character, having degree 9. Since there are no elements of order 15, Sylow's theorem and the Burnside transfer theorem now show that the normaliser of a Sylow 5-subgroup is a dihedral group of order 10. Hence there are two conjugacy classes of elements of order 5 and one each of elements of orders 2 and 4. In addition, the centraliser of an involution must now have order 8 and, since a group with Sylow 2-subgroup $Z_4\times Z_2$ cannot be simple by Burnside's transfer theorem, the centraliser of an element of order 4 has order 4. (In fact, a Sylow 2-subgroup must be dihedral.) Labelling conjugacy classes by the orders of elements, we may now compute the entire character table of G (Table 2.5). This can be done using the orthogonality relations together with the observation that, if y is a real element of order 4 and χ is an irreducible character, then $\chi(y)\in\mathbb{Z}$ and $\chi(y)\equiv\chi(1)\pmod 2$. The character values α and β on the elements of order 5 are $-(\omega+\omega^{-1})$ and $-(\omega^2+\omega^{-2})$ where ω is a primitive fifth root of 1.

Table 2.5

	1	2	4	3_1	3_2	5_1	5_2
1_G	1	1	1	1	1	1	1
χ_2	8	2	0	-1	-1	α	β
χ_3	8	0	0	-1	-1	β	α
χ_4	10	-2	0	1	1	0	0
χ_5	5	1	-1	2	-1	0	0
χ_6	5	1	-1	-1	2	0	0
χ_7	9	1	1	0	0	-1	-1

Although we now have the character table of G, we have not yet established the claimed isomorphism. Indeed, as we may compare with the situation that arose in connection with Theorem 44, we cannot expect character theory to achieve this by itself since we can essentially obtain only numerical information by these methods. In this case, however, it is clear that the major step will be to show the existence of a subgroup of index 6, namely of order 60.

To do this, we first make a digression.

Proposition 54. *The following are presentations for the groups A_4, S_4 and A_5.*

(i) $A_4 \cong \langle a, b \mid a^2 = b^3 = (ab)^3 = 1 \rangle$.
(ii) $S_4 \cong \langle a, b \mid a^2 = b^3 = (ab)^4 = 1 \rangle$.
(iii) $A_5 \cong \langle a, b \mid a^2 = b^3 = (ab)^5 = 1 \rangle$.

The proof of this result involves the study of tesselation groups and lies outside the scope of this book (although it is not too deep). Since, in the groups in question, the elements actually have the orders specified by the presentation but this is not the case in any proper homomorphic image, we have the following as an immediate consequence.

Corollary 55. *Let G be a finite group and let x and y be elements of orders 2 and 3. If their product xy has order 3, 4 or 5, then $\langle x, y \rangle \cong A_4$, S_4 or A_5 respectively.*

The presence of such elements can be detected by the nonvanishing of a suitable class algebra constant, and this in turn can be determined from the character table.

We can now complete the identification of G as A_6 in Example 3. Let

$y \in 5_1$. Then

$$|\mathscr{A}(2, 3_1; y)| = \frac{|G|}{72} = 5.$$

So G contains subgroups isomorphic to A_5 and it is then a trivial exercise to show that $G \cong A_6$.

We remark that the only groups which satisfy the hypothesis of Example 2 are PSL(2, 5) and PSL(2, 7). The arguments used to establish this are based on methods similar to the above, although they are more complex since part (ii) of Proposition 53 is not available. The characterisation of these groups by this property was first obtained by Feit and Thompson (1962). The argument given in Example 2 leads to an equation

$$\frac{|G|}{|C(\tau)|} \cdot \frac{(x-u)^2}{x(x+1)} = 3 \tag{9.7}$$

in the notation established there. The further character theoretic formulae can be obtained by using Proposition 53. An alternative treatment using Corollary 55 has been given by G. Higman (1968). (See Exercise 2.)

Exercises

1. In Example 3 above, show that $\mathscr{A}(2, 3_2; y) \neq \emptyset$. Deduce that S_6 possesses an outer automorphism of order 2.
2. Let G be a group satisfying the hypothesis of Example 2.

   (i) Establish (9.7).

   (ii) By using Corollary 55, show that G possesses subgroups isomorphic to A_4, and also to S_4 or A_5 if G possesses elements of orders 4 or 5 respectively.

   [Remark. This enables one to study 2-subgroups of G which are maximal with respect to being normalised by an element of order 3.]

3. Let G be a group and, in the usual notation, define nonnegative integers a_{ijkl} by

$$C_i C_j C_k = \sum_{l=1}^{r} a_{ijkl} C_l.$$

   Obtain a character theoretic formula for a_{ijkl}.
   Show also, for any i, j, k, l, that

$$\sum_m a_{ijm} a_{mkl} = \sum_m a_{jkm} a_{iml}.$$

4. Let g be an element in a group G and let $c(g)$ denote the number of ordered pairs of involutions (τ, τ') such that $g = \tau\tau'$. If $\tau_1, \ldots, \tau_n$ are

representatives of the conjugacy classes of involutions in G, show that

$$c(g) = |G| \sum_{i=1}^{r} \frac{b_i \chi_i(g)}{\chi_i(1)}$$

where

$$b_i = \left(\sum_{i=1}^{n} \frac{\chi_i(\tau_j)}{|C(\tau_j)|} \right)^2.$$

[Hint. Let T be the sum (in $\mathbb{C}G$) of the involutions of G and consider T^2.]

5. Using the character tables for the groups PSL$(2, q)$ computed in Exercise 1 of Section 7, show that PSL$(2, q)$ possesses subgroups isomorphic to S_4 when $q \equiv \pm 1 \pmod 8$ and to A_5 when $q \equiv \pm 1 \pmod 5$.
   Deduce that SL(2, 5) can occur as a Frobenius complement in some Frobenius group.

10 A characterisation of the groups SL$(2, 2^n)$

This section will be devoted to a discussion of the following special case of a fundamental theorem due to Brauer, Suzuki and Wall (1958).

Theorem 56.[†] *Let G be a finite simple group having an elementary abelian Sylow 2-subgroup T of order 2^n, $n \geqslant 2$. Suppose that $C_G(\tau) = T$ for all $\tau \in T^\#$ Then $G \simeq$ SL$(2, 2^n)$.*

The smallest case, $n = 2$, was considered in Example 1 of the previous section but, whilst there it was sufficient to consider just involutions, here we shall need to examine the structure of the group in greater detail. As in that case, we shall study the values taken by class algebra constants but here we shall employ character theory to do two things: first we will show that every element in the group is real, and then we will analyse character values and degrees to determine the group order.

The Brauer–Suzuki–Wall theorem was the first characterisation of a class of simple groups of even order in terms of a 'local' group theoretic hypothesis, and it has played an important role in the classification of finite simple groups. They actually proved a characterisation of all the groups PSL$(2, q)$, that for q even being under weaker hypotheses than those of Theorem 56. The special case that we shall consider may be viewed an extension of the discussion of the previous section; the full result would need the methods of the next chapter. In fact, we shall not

[†] This case was also discovered by Burnside (1900) in a little known paper to which he seems never to have referred again. His hypothesis, equivalent to ours, is that every element in a simple group G of even order has either odd order or order 2. His proof, which is character-free, also contains the Brauer–Fowler lemma, whose significance he probably did not recognise. I am grateful to Walter Feit for bringing this to my attention.

give the complete proof but refer the interested reader to the original paper. We should, however, mention that a proof independent of character theory (but still by counting involutions) has been given by Bender (1974); see also the account given by Suzuki [S II; pp. 497ff.]

Throughout this section, we shall assume the hypothesis of Theorem 56 and the standard notation of this chapter. Fix $\tau \in T^{\#}$ and suppose that y is an involution in $G - T$ which is not conjugate to τ. Then τy has even order and, if z is the involution in $\langle \tau y \rangle$, it follows that $z \in C(\tau) \subseteq T$ and hence $y \in C(z) \subseteq T$. This contradiction establishes the following.

Lemma 57. *All involutions in G are conjugate.*

By Burnside's lemma (Lemma 1.47), all involutions in T are conjugate in $N(T)$. Hence $|N(T)| = 2^n(2^n - 1)$. Also, since $\langle T, T^g \rangle \subseteq C(T \cap T^g)$ for any $g \in G$, it follows that $T \cap T^g = 1$ if $g \notin N(T)$. Thus $[G:N(T)] \equiv 1 \pmod{2^n}$ and hence, for some positive integer N,

$$|G| = (2^n N + 1)2^n(2^n - 1) \geqslant (2^n + 1)2^n(2^n - 1). \tag{10.1}$$

Now any nonidentity element of G is either an involution or of odd order. So any real element is necessarily strongly real. We shall label the conjugacy classes in such a way that $\mathscr{C}_1 = \{1\}$, $\mathscr{C}_2 = \{\text{involutions}\}$, and $\mathscr{C}_3, \ldots, \mathscr{C}_s$ are the conjugacy classes of nonidentity real elements of odd order. If $j = 3, \ldots, s$ and $g_j \in \mathscr{C}_j$, then $C^*(g_j)$ is a Frobenius group in which every element of $C^*(g_j) - C(g_j)$ is an involution, and hence

$$a_{22j} = |C(g_j)| \quad \text{for } j = 3, \ldots, s.$$

Also, $\mathscr{A}(\mathscr{C}_2, \mathscr{C}_2; \tau) = \mathscr{A}_T(\mathscr{C}_2, \mathscr{C}_2; \tau)$: hence

$$a_{222} = 2^n - 2. \tag{10.2}$$

Thus

$$C_2^2 = \frac{|G|}{2^a} C_1 + (2^n - 2)C_2 + \sum_{j=3}^{s} |C(g_j)| C_j.$$

Counting the number of elements involved in this expression, we see that

$$|G| = (s-1)2^{2n} - 2^n. \tag{10.3}$$

We now employ character theory to show that *every* class is real. Let $\chi_1 = 1, \chi_2, \ldots, \chi_r$ be the irreducible characters of G and put

$$f_i = \chi_i(1), \quad z_i = \chi_i(\tau);$$

we shall order the characters so that

(a) $z_i > 0$ for $i = 1, \ldots, p$,
(b) $z_i < 0$ for $i = p+1, \ldots, p+q$,
(c) $z_i = 0$ for $i = p+q+1, \ldots, r$.

Since G is simple, we can note that $|z_i| < f_i$ for $i \geqslant 2$.

The multiplicity $(\chi_i|_T, 1)_T$ is a nonnegative integer b_i; hence, for each i,

$$f_i + (2^n - 1)z_i = b_i \cdot 2^n$$

and so we may write

$$f_i = z_i + 2^n c_i \tag{10.4}$$

where $c_i = b_i - z_i$. For the three cases (a), (b) and (c), we have

$$c_i \geqslant 1, \quad c_i \geqslant |z_i|, \quad c_i \geqslant 1$$

respectively.

We shall repeatedly apply the degree formula

$$|G| = \sum_i f_i^2. \tag{10.5}$$

Since $r \geqslant s$, this together with (10.3) implies that $f_i < 2^n$ for some i; in view of (10.4), this can occur only in case (b). Also $\sum_i z_i^2 = 2^n$: hence $|z_i| < 2^n$ so that $f_i < 2^n$ only if $c_i = 1$ and $z_i = -1$. Thus,

$$\text{if } f_i < 2^n \quad \text{then} \quad f_i = 2^n - 1. \tag{10.6}$$

Suppose that there are b characters of degree $2^n - 1$: then, since $z_i \neq 0$ for such characters,

$$b \leqslant 2^n - 1. \tag{10.7}$$

The degree formula (10.5) now yields the inequality

$$|G| \geqslant 1 + b(2^n - 1)^2 + (r - b - 1)2^{2n}$$

which, by (10.3), becomes

$$(r - s)2^{2n} \leqslant (2^{n+1} + 1)b - (2^n + 1):$$

in turn, with (10.7), this yields the inequality

$$(r - s)2^{2n} \leqslant 2^{2n+1} - (2^{n+1} + 2).$$

Since nonreal classes occur in pairs, $r - s$ must be even, and this then implies that $r = s$; so we have shown that all conjugacy classes in G are real.

If any character had degree as large as $2^{n+1} - 2$ then, as $r = s$, the formulae (10.3), (10.5) and the inequality (10.7) would yield a contradiction. So, since $|z_i| \leqslant 2^n$, (10.4) and (10.6) force, for the three cases,

(a) $c_i = 1$,
(b) $c_i = 1, \quad z_i = -1$,
(c) $c_i = 1$,

and, in turn, these give the degrees and character values

(a) $f_i = 1, \quad f_i = z_i + 2^n, \quad z_i > 0 \quad \text{for } i = 2, \ldots, p$,

(b) $f_i = 2^n - 1, \quad z_i = -1$ for $i = p+1, \ldots, p+q$,
(c) $f_i = 2^n, \quad z_i = 0$ for $i = p+q+1, \ldots, r$.

In particular, $q = b$.

The orthogonality relations give equations

$$\sum_{i=1}^{r} z_i^2 = 1 + \sum_{i=2}^{p} z_i^2 + q = 2^n$$

and

$$\sum_{i=1}^{r} f_i z_i = 1 + \sum_{i=2}^{p} (z_i^2 + 2^n z_i) - q(2^n - 1) = 0,$$

whence

$$\sum_{i=2}^{p} z_i = q - 1.$$

It follows that

$$q = 1 + \sum_{i=2}^{p} z_i \leqslant 1 + \sum_{i=2}^{p} z_i^2 = 2^n - q \tag{10.8}$$

and, in particular, that

$$2q \leqslant 2^n. \tag{10.9}$$

We now compute the class algebra constant a_{222} using the character values and relations above. Together with (10.2), we can obtain the equation

$$\frac{|G|}{2^{2n}}\left(1 + \sum_{i=2}^{p} \frac{z_i^3}{2^n + z_i} - \frac{q}{2^n - 1}\right) = 2\mathrm{n} - 2. \tag{10.10}$$

Now $z_i \geqslant 1$ if $i = 2, \ldots, p$, and so

$$\frac{z_i^3}{2^n + z_i} \geqslant \frac{z_i^2}{2^n + 1}.$$

Substituting this into (10.10) and combining with the second equality of (10.8), we obtain the further inequality

$$(2^n - 2)2^{2n}(2^n + 1)(2^n - 1) \geqslant |G|(2^{2n+1} - 2^{n+1}q - 2^{n+1})$$

which, combined with the original inequality (10.1), then yields

$$2^n - 2 \geqslant 2^{n+1} - 2q - 2.$$

Thus $2q \geqslant 2^n$ and hence, in view of the inequality (10.9), $q = 2^{n-1}$. But this now implies also that the inequalities (10.1) and (10.8) are actually equalities, and hence

$$|G| = (2^n + 1)2^n(2^n - 1),$$

$$z_1 = z_2 = \cdots = z_p = 1,$$

$$p = q;$$

finally, from the degree equation (10.5), we deduce that G has exactly one character χ_r of degree 2^n.

As with the examples in the previous section, it still remains to identify G as SL$(2, 2^n)$, and again we shall appeal to Zassenhaus' theorem (Proposition 45) by showing that G has a permutation representation of degree $2^n + 1$ in which it acts as a sharply triply transitive group. We shall attack this directly; another approach, using the character theoretic information already available, will be given in Exercise 2.

We represent G as a permutation group of degree $2^n + 1$ via conjugation on its Sylow 2-subgroups. Certainly the action is transitive with $N(T)$ the stabiliser of T, but then it is doubly transitive since no element of $T^{\#}$ fixes any other Sylow 2-subgroup. Let H be the subgroup of $N(T)$ fixing some other Sylow 2-subgroup. The action of $N(T)$ on the 2^n Sylow 2-subgroups of G other than T is equivalent to its action on the cosets of H and, by Proposition 35, $N(T)$ is a Frobenius group with Frobenius kernel T and complement H. So the action of $N(T)$ is the natural one of a Frobenius group, with every element of $H^{\#}$ fixing exactly one Sylow 2-subgroup other than T, and $|H| = 2^n - 1$. Hence $N(T)$ acts 2-transitively on the cosets of H, and G acts 3-transitively on the conjugates of T with no nonidentity element fixing more than two points: that is, G acts as a sharply triply transitive permutation group.

Exercises

(Throughout, G is a group which satisfies the hypotheses of Theorem 56.)

1. Let g be a nonidentity element of odd order in G. From an assumption that all such elements are strongly real, establish the following.

   (i) $C(g)$ is an abelian group.

   (ii) Let p be an odd prime. Then a Sylow p-subgroup P of G is abelian, and $N(P)$ has order divisibly by 2 but not 4.

   (iii) $N(P)$ has an abelian subgroup of index 2 which is a Hall subgroup of G.

   (iv) Every subgroup of G of odd order is abelian.

2. Let H be defined as in the completion of the proof of Theorem 56. Show that every element of G of order dividing $2^n - 1$ is conjugate to an element of H. Show further that, if g is a nonidentity element of odd order, then $\chi_r(g) \geqslant 1$ if g has order dividing $2^n - 1$ while $\chi_r(g) = -1$ otherwise. (χ_r is the character of degree 2^n.)

   By counting elements and examining the relation $\sum |\chi_r(x)|^2 = |G|$, deduce that $\chi_r(g) = 1$ in the first case above, and deduce that the action of G is sharply triply transitive.

11 Rigidity in finite groups

We shall conclude this chapter with a recent development.

A longstanding question in group theory is the following, the so-called Inverse Galois Problem. *Is every finite group the Galois group of some polynomial over the rationals* $\mathbb{Q}$? This has been known to be the case for (soluble) groups of odd order for some time.† More recently, Belyi (1980) studied the question for Galois extensions of the maximal abelian extension of $\mathbb{Q}$ and Thompson (1984) has introduced a concept of rigidity in finite groups which can be used to show that certain groups do occur as Galois groups over a field of character values. The derivation of these theorems is outside the scope of this book, relying on some deep ideas from algebraic geometry; what is relevant for us is that knowledge of character tables can contribute to the verification of the conditions required.

For $k \geqslant 3$, let $\mathscr{C}_1, \ldots, \mathscr{C}_k$ be an ordered set of conjugacy classes in a finite group G, not necessarily distinct. Let

$$\mathscr{A} = \mathscr{A}(\mathscr{C}_1, \ldots, \mathscr{C}_k) = \{(x_1, \ldots, x_k) | x_i \in \mathscr{C}_i, x_1 \cdots x_k = 1\}.$$

Then we say that $\mathscr{A}$ is *rigid* if $\mathscr{A} \neq \varnothing$ and

(i) G permutes the members of $\mathscr{A}$ transitively under conjugation, and
(ii) $G = \langle x_1, \ldots, x_k \rangle$ for every $(x_1, \ldots, x_k) \in \mathscr{A}$.

Further, G is *rigid* if G contains conjugacy classes $\mathscr{C}_1, \ldots, \mathscr{C}_k$ such that $\mathscr{A}(\mathscr{C}_1, \ldots, \mathscr{C}_k)$ is rigid.

A conjugacy class $\mathscr{C}$ in G is *rational* if $\mathscr{C}$ contains every generator of $\langle x \rangle$ for $x \in \mathscr{C}$. Then G is *rationally rigid* if G is rigid with respect to a collection of rational conjugacy classes. We can now state Thompson's theorem or, more precisely, a special case of it.‡

Proposition 58. *Let G be a group with $Z(G) = 1$, and suppose that G is rationally rigid. Then G is the Galois group of a finite normal extension of the rationals over* $\mathbb{Q}$.

In order to see how the conditions may be verified, we first note that a conjugacy class is rational if and only if every character takes rational values on its elements. (Cf. Lemma 15 and Theorem 16, and also Exercise 30 of Section 1.) So the rationality of a conjugacy class may be verified from the character table. Next, we observe that $\mathscr{A}$ is rigid if and only if

(i)′ $|\mathscr{A}| = |G|$, and
(ii)′ $G = \langle x_1, \ldots, x_k |$ some $x_i \in \mathscr{C}_i$, $x_1 \cdots x_k = 1 \rangle$.

† This theorem is due to Shafaravich. Originally the theorem was claimed for soluble groups, but a certain result on which his proof relies applies only to groups of odd order.
‡ Thompson also required that $k \leqslant 6$, but this is apparently unnecessary. See Feit (1984).

Table 2.6

1^5	$1^3{\cdot}2$	$1{\cdot}2^2$	$1{\cdot}4$	$1^2{\cdot}3$	$2{\cdot}3$	5
1	1	1	1	1	1	1
4	2	0	0	1	−1	−1
5	−1	1	1	−1	−1	0
6	0	−2	0	0	0	1
1	−1	1	−1	1	−1	1
4	−2	0	0	1	1	−1
5	1	1	−1	−1	1	0

The form of the set $\mathscr{A}$ should be highly reminiscent of the class algebra constant calculations of the preceding sections, except that every $x_k \in \mathscr{C}_k$ is taken: also, $x_k^{-1} = x_1 \cdots x_{k-1}$. An easy extension of the proof of Corollary 9 gives a formula

$$|\mathscr{A}(\mathscr{C}_1, \ldots, \mathscr{C}_k)| = \frac{|G|^{k-1}}{|C(x_1)| \cdots |C(x_k)|} \sum \frac{\chi(x_1) \cdots \chi(x_k)}{\chi(1)^{k-2}} \tag{11.1}$$

where $x_i \in \mathscr{C}_i$ and the summation is carried over all irreducible characters of G. So condition (i)′ may be checked from the character table of G. Condition (ii)′ is more difficult. (See the remark at the end of the section.) Either it must be verified directly, or else the process repeated to show that, for each maximal subgroup H of G and $x_i \in \mathscr{C}_i \cap H$, the product $x_1 \cdots x_k$ cannot be 1; this also can be checked by showing that all the relevant sums for (11.1) are zero when computed in such subgroups.

As an example, we consider the symmetric group S_5. Of course, it is well-known that this is a Galois group, but we want an easily accessible example to demonstrate the method. If we label the conjugacy classes of S_5 by their permutation types, we have the character table 2.6.

Certainly, every conjugacy class is rational. It is easy to check that a 4-cycle x and a 5-cycle y either generate the whole of S_5 or a Frobenius group of order 20. Hence, if τ is a transposition and $\tau x = y$, we have $S_5 = \langle \tau, x \rangle$. Only two characters are nonzero on all three classes and the formula (11.1) yields

$$|\mathscr{A}(1^3{\cdot}2, 1{\cdot}4; 5)| = 120;$$

hence S_5 is rationally rigid with respect to the classes $1^3{\cdot}2$, $1{\cdot}4$ and 5. We observe, in passing, the additional factor $|G|/|C(x_3)|$ in the formula as compared to that for the class algebra constant which, in this case, is 5. This is because we do not fix the member of the third class in carrying out the computation.

In fact, a similar argument shows that all the symmetric groups are rationally rigid. An analysis of the characters of the symmetric groups will not be given in this book (see, for example, James and Kerber [JK]), but it is the case that, for the classes $1^{n-1}\cdot 2, 1\cdot(n-1)$ and n in S_n, the only characters which are simultaneously nonzero on $1\cdot(n-1)$ and n are the two linear characters. So for these three classes one sees that $|\mathscr{A}| = |G|$, while the generation condition is easily checked.

In Thompson's method, simple groups play a central role in a programme to settle the inverse Galois problem. However, it is not always possible to show that a simple group is in fact rigid. An observation due to Serre can be applied. If a group G is rationally rigid for $k=3$ and G contains a subgroup G_0 of index 2, then G_0 is also a Galois group of some finite normal extension of the rationals over $\mathbb{Q}$. Hence, in particular, the alternating groups also occur as Galois groups over $\mathbb{Q}$. This argument has also been applied in other cases.

As a final remark, we observe that, for $k=3$, Hunt (1986) has defined a *candidate triple* as a set of three rational conjugacy classes for which an examination of the character table showed that condition (i)′ holds and, accordingly, he examined many of the sporadic simple groups using a computer. In the case of the fourth Janko group J_4, fourteen candidate triples were obtained; all but one were easily eliminated, but the last, corresponding to particular classes of elements of orders 2, 4 and 11, was shown to fail only with considerable difficulty. There is a corresponding nonzero class algebra constant in the Mathieu group M_{24}. The group J_4 has a maximal subgroup M having a normal subgroup H of order 2^{11} with $M/H \cong M_{24}$, but deciding whether a triple of elements in the quotient could be lifted to elements of M of the correct orders and lying in the appropriate conjugacy classes proved particularly difficult.

Exercises

1. Prove the formula (11.1).
2. Prove that the group PGL(2.7) is rationally rigid with respect to classes of elements of orders 2, 6 and 7 where the involutions are the cubes of elements of order 6. Deduce that PSL(2, 7) is a Galois group over $\mathbb{Q}$.

   [The character table is given in Table 2.2. Note that the Sylow 7-normaliser is a Frobenius group of order 42, maximal in PGL(2, 7).]

3

Suzuki's theory of exceptional characters

One of the major questions that a finite group theorist often wants to answer is the following.

Given a group G which contains a subgroup H of known isomorphism type, what can be deduced about the group G?

In terms of character theory, one can ask how knowledge of the characters of H can influence the characters of G. We have seen this line of attack pursued successfully when we proved Frobenius' theorem in Chapter 2. Indeed, we noted that a careful examination of the proof would reveal that a portion of the character table lifted directly into the character table of G. Such a situation requires two basic ingredients. Firstly, there must be a one-to-one correspondence between certain conjugacy classes in H and certain conjugacy classes in G – in particular, no two of these conjugacy classes of H may fuse in G – and, secondly, there should correspond to certain irreducible characters of H irreducible characters of G which should take the same, or related, values on the distinguished conjugacy classes. We shall not give a precise definition here, but in the language of Suzuki these characters are called *exceptional characters.*

In this chapter, we shall follow the approach taken by Suzuki (1959); however, whereas Suzuki studied the usual character ring $\mathscr{C}\mathscr{h}(G)$, we shall take a larger coefficient ring to start with since his results extend verbatim. Having established the basic machinery, we will then give some applications where 'best possible' results can be obtained. Unfortunately, this can not always be achieved by these methods alone, and in the next chapter we will follow an approach due to Feit, as improved by Sibley.

1 Closed and special subsets

We start with a slight generalisation of the usual character ring. Let H be a group. Let K be an algebraic number field which is a splitting field for H and all of its subgroups, and let $\mathscr{C}(H, K)$ denote the K-algebra of K-valued class functions on H. Then $\mathscr{C}(H, K)$ has essentially the same properties as $\mathscr{C}(H, \mathbb{C})$ since it contains the irreducible (complex) characters as a basis. Let R be a subring of K. In place of the usual character ring, we consider the R-lattice in $\mathscr{C}(H, K)$ consisting of all R-linear combinations of the irreducible characters of H, that is

$$\mathscr{Ch}_R(H) = R \otimes_{\mathbb{Z}} \mathscr{Ch}(H).$$

If C is a normal subset of H, let

$$\mathscr{M}(C) = \{\varphi \in \mathscr{Ch}_R(H) \mid \varphi(g) = 0 \text{ if } g \notin C\},$$
$$\hat{\mathscr{M}}(C) = \{\psi \in \mathscr{Ch}_R(H) \mid (\varphi, \psi)_H = 0 \text{ for all } \varphi \in \mathscr{M}(C)\},$$

and

$$\hat{C} = H - C.$$

Throughout this section, we shall assume that R is chosen to be a principal ideal domain. (In Chapters 5 and 6, we shall drop this restriction and, for example, take R as the ring of algebraic integers in K.)

Suppose that C consists of t conjugacy classes of H. The R-module $\mathscr{Ch}_R(H)$ is certainly free and finitely generated; since R is a principal ideal domain, $\mathscr{M}(C)$ is a finitely generated free submodule of $\mathscr{Ch}_R(H)$. We may establish the following basic properties.

Proposition 1. (i) *$\mathscr{M}(C)$ is a direct summand of $\mathscr{Ch}_R(H)$ of rank at most t. An R-basis of $\mathscr{M}(C)$ is linearly independent over K.*

(ii) *$\mathscr{M}(C)$ is an ideal of $\mathscr{Ch}_R(H)$.*

Proof. There exists an R-basis $\{\varphi_1, \ldots, \varphi_n\}$ for $\mathscr{Ch}_R(H)$ such that, for some $m \leqslant n$ and nonzero $c_1, \ldots, c_m \in R$, the class functions $c_1\varphi_1, \ldots, c_m\varphi_m$ form an R-basis for $\mathscr{M}(C)$. If $c_i\varphi_i \in \mathscr{M}(C)$, then clearly $\varphi_i \in \mathscr{M}(C)$ so that $\{\varphi_1, \ldots, \varphi_m\}$ is an R-basis for $\mathscr{M}(C)$. Hence $\mathscr{M}(C)$ is a direct summand of $\mathscr{Ch}_R(H)$. Furthermore, since $\mathscr{Ch}_R(H)$ has an R-basis which is linearly independent over K, namely the irreducible characters of H, and the basis $\{\varphi_1, \ldots, \varphi_n\}$ can be obtained from it by a unimodular transformation over R, this basis is linearly independent over K. This proves the assertion about the rank of $\mathscr{M}(C)$ also.

Since products of characters are characters, part (ii) is immediately clear.

We shall want to restrict our attention to certain unions of conjugacy classes of elements of H, for which Proposition 2 will provide the necessary character theoretic motivation.

Proposition 2. *The following conditions are equivalent for $g, h \in H$.*

(i) *Whenever $\varphi \in \mathscr{Ch}(H)$ and $\varphi(g) = 0$, then $\varphi(h) = 0$.*
(ii) *The elements g and h generate conjugate cyclic subgroups of H.*

Proof. First suppose that condition (ii) holds. If g has order m, then we may assume that $h = g^s$ where $(s, m) = 1$. If $\varphi(g) = 0$, then

$$\varphi(g) = \sum_i \varepsilon_i \omega_i = 0$$

where, for each i, ω_i is an mth root of unity and $\varepsilon_i = \pm 1$. So

$$\varphi(h) = \sum_i \varepsilon_i \omega_i^s.$$

Since $(s, m) = 1$, the map $\omega \to \omega^s$ defines an automorphism of $\mathbb{Q}(\omega)$ when ω is a primitive mth root of unity: thus $\varphi(h) = 0$.

In order to prove the converse, we need a lemma.

Lemma 3. *Let S be a cyclic group. Then there is a generalised character θ of S with $(\theta, 1)_S = 1$, which vanishes on every proper subgroup of S.*

Proof. Let $S = S_1 \times \cdots \times S_k$ be the decomposition of S as a direct product of its Sylow subgroups. For each $i = 1, \ldots, k$, let T_i be the maximal subgroup of S_i, and let λ_i be a nonprincipal character of S with kernel $T_i \times \prod_{j \neq i} S_j$. Let $\lambda_0 = 1_S$. Then it is easily verified that the product

$$\theta = \prod_{i=1}^{k} (\lambda_0 - \lambda_i)$$

has the desired properties since any nontrivial product of distinct characters λ_i (for $i \geqslant 1$) is a nonprincipal irreducible character of S.

We may now complete the proof of Proposition 2. Assume that (ii) is false. Construct θ as in Lemma 3 for the subgroup $S = \langle h \rangle$, noting that θ vanishes on the nongenerators of S. By the Frobenius reciprocity theorem,

$$(\theta^H, 1)_H = (\theta, 1)_S = 1. \tag{1.1}$$

Now θ^H vanishes on every element of H not conjugate to a generator of S; in particular, $\theta^H(g) = 0$. If (i) holds, then $\theta^H(h) = 0$ and thus $\theta^H(h^t) = 0$ whenever $\langle h^t \rangle = S$, since (ii) implies (i). But then $\theta^H \equiv 0$, contrary to (1.1). So condition (i) implies condition (ii).

The conditions of Proposition 2 impose an equivalence relation on the elements of H, motivating the following definition.

Definition. A normal subset C of H is *closed* if, whenever $g \in C$, every generator of $\langle g \rangle$ lies in C also.

In this case, we can strengthen the bound of Proposition 1 to an equality (but see also Exercise 7). First we need two lemmas.

Lemma 4. *Suppose that C is a closed subset of H. Define a function $\theta \in \mathscr{C}(H, K)$ by*

$$\theta(h) = \begin{cases} 0 & \text{if } h \notin C, \\ |H| & \text{if } h \in C. \end{cases}$$

Then $\theta \in \mathscr{C}\mathscr{h}(H)$.

Proof. Let χ be an irreducible character of H. Then

$$(\chi, \theta) = |H|^{-1} \sum_{h \in H} \overline{\theta(h)} \chi(h) = \sum_{h \in C} \chi(h).$$

We may assume, without loss, that K is a normal extension of $\mathbb{Q}$ containing a primitive $|H|$th root ε of unity. Then any automorphism α of K maps $\varepsilon \to \varepsilon^m$ for some m coprime to $|H|$, and α: $\chi(h) \to \chi(h^m)$. Since C is closed, (χ, θ) is invariant under α and hence rational. As (χ, θ) is an algebraic integer, we see that $(\chi, \theta) \in \mathbb{Z}$. (Compare this argument with the proof of Theorem 2.16.)

Lemma 5. *If C_1 and C_2 are disjoint closed subsets of H, then $\mathscr{M}(C_1 \cup C_2)/(\mathscr{M}(C_1) + \mathscr{M}(C_2))$ is a torsion module.*

Proof. Certainly $\mathscr{M}(C_1) + \mathscr{M}(C_2) \subseteq \mathscr{M}(C_1 \cup C_2)$. Now, if $\varphi \in \mathscr{M}(C_1 \cup C_2)$, it suffices to show that $|H|\varphi \in \mathscr{M}(C_1) + \mathscr{M}(C_2)$. If θ_1 and θ_2 are constructed for C_1 and C_2 respectively as in Lemma 4, then

$$|H|\varphi = \theta_1\varphi + \theta_2\varphi \in \mathscr{M}(C_1) + \mathscr{M}(C_2),$$

since clearly $\theta_i\varphi \in \mathscr{M}(C_i)$.

For a submodule $\mathscr{M}$ of $\mathscr{C}\mathscr{h}_R(H)$, let $r(\mathscr{M})$ denote its rank over R. Then we may immediately conclude the following.

Corollary 6. $r(\mathscr{M}(C_1 \cup C_2)) = r(\mathscr{M}(C_1)) + r(\mathscr{M}(C_2))$.

We can now complete the determination of $r(\mathscr{M}(C))$.

Theorem 7. *If C is a closed subset of the group H consisting of t conjugacy classes, then $r(\mathscr{M}(C)) = t$. Furthermore, $\hat{\mathscr{M}}(C) = \mathscr{M}(\hat{C})$.*

Proof. Since $H = C \cup \hat{C}$, the first statement follows from the bound of Proposition 1 and Corollary 6. To see the second claim, let L be the quotient field of R. Then we need only observe from Lemma 5 that

$$L \otimes_R (\mathscr{M}(C) \oplus \mathscr{M}(\hat{C})) = L \otimes_R \mathscr{C}\mathscr{h}_R(H),$$

whence

$$K \otimes_R \mathscr{M}(C) \oplus K \otimes_R \mathscr{M}(\hat{C}) = \mathscr{C}(H, K),$$

so that both $\hat{\mathcal{M}}(C)$ and $\mathcal{M}(\hat{C})$ consist precisely of those elements of $\mathcal{C}h_R(H)$ which lie in $K \otimes_R \mathcal{M}(\hat{C})$.

We shall now link these ideas with those that arise out of the proof of Frobenius' theorem. For the remainder of this section and the next, we shall be concerned with a subgroup H of a group G. We shall be interested, in particular, in generalised characters of H which vanish outside certain conjugacy classes: in view of Proposition 2, we can assume that those classes form a closed set in H. To avoid ambiguity, we shall write $\mathcal{M}_H(C)$ instead of $\mathcal{M}(C)$, and the ring R will be taken to be the ring of integers $\mathbb{Z}$.

Definition. Let G be a group, H a subgroup of G, and S a normal subset of H. Then S is a *set of special classes* if

(i) whenever x and y are elements of S which are conjugate in G, then x and y are conjugate in H, and
(ii) if $x \in S$, then $C_G(x) \subseteq H$.

We note that, if $H \neq G$, then $1 \notin S$. Furthermore, if $x, y \in S$ and $x^g = y$, then necessarily $g \in H$. In the proof of Frobenius' theorem, if H is the stabiliser of a point, then the set of nonidentity elements of H forms a set of special classes.

An equivalent definition can be given by the following result whose proof is left as an exercise. (Cf. the proof of Frobenius' theorem.)

Proposition 8. *Let H be a subgroup of a group G, and let S be a normal subset of H. Then S forms a set of special classes if and only if $S \cap S^g = \varnothing$ whenever $g \in G - H$.*

Similarly, the basic character theoretic concepts arising out of the induction map in the proof of Frobenius' theorem extend to the situation where special classes exist.

Theorem 9. *Let H be a subgroup of a group G and let S be a set of special classes in H. Let θ be a function in $\mathcal{C}(H, \mathbb{C})$ which vanishes outside S. Then the following hold.*

(i) $\theta^G(g) = \theta(g)$ *if* $g \in S$.
(ii) $\theta^G(g) = 0$ *if g is an element of G which is not conjugate to any element of S.*
(iii) *If φ is a function in $\mathcal{C}(H, \mathbb{C})$ which vanishes on any element of $H - S$ which is conjugate to an element of S, then $(\theta^G, \varphi^G)_G = (\theta, \varphi)_H$.*

Proof. (i) and (ii) are straightforward applications of the formula for induced class functions (Theorem 2.30). To see (iii), we observe that the condition on φ is sufficient to ensure that $\varphi^G(g) = \varphi(g)$ whenever $g \in S$, so that

$$\begin{aligned}(\theta^G, \varphi^G)_G &= |G|^{-1} \sum_{g \in G} \theta^G(g)\overline{\varphi^G(g)} \\ &= |G|^{-1}[G:H] \sum_{g \in S} \theta(g)\overline{\varphi(g)} \\ &= |H|^{-1} \sum_{g \in H} \theta(g)\overline{\varphi(g)} = (\theta, \varphi)_H.\end{aligned}$$

We note that, if in addition S is a closed subset of special classes in H so that the earlier results on $\mathscr{M}_H(S)$ apply, then (iii) tells us that induction is an *isometry* from $\mathscr{M}_H(S)$ to $\mathscr{C}\mathscr{h}(G)$. This concept will be central to our considerations in this chapter and the next. In the next section we shall derive an algorithm for analysing this isometry, but we conclude this section by describing a situation in which the isometry leads to a natural correspondence between certain irreducible characters of H and irreducible characters of G. We shall first give this in a form sufficient for many applications, and then, in Theorem 11, in a more precise form as originally exploited by Brauer and Suzuki. (See Suzuki (1955).)

Theorem 10. *Let H be a subgroup of the group G and let S be a set of special classes in H. Suppose that H has s irreducible characters $\varphi_1, \ldots, \varphi_s$ such that $\varphi_i(h) = \varphi_j(h)$ whenever $i \neq j$ and $h \in H - S$. If $s \geqslant 2$, then there exist a sign $\varepsilon = \pm 1$ and irreducible characters $\chi^{(1)}, \ldots, \chi^{(s)}$ of G such that, for $1 \leqslant i, j \leqslant s$,*

$$(\varphi_i - \varphi_j)^G = \varepsilon(\chi^{(i)} - \chi^{(j)}).$$

If $s \geqslant 3$, then ε is uniquely determined.

If $s \geqslant 2$ and χ is an irreducible character of G distinct from $\chi^{(1)}, \ldots, \chi^{(s)}$, then $\chi|_H$ contains each φ_i with the same multiplicity.

Proof. For $i = 1, \ldots, s-1$, let $\psi_i = \varphi_i - \varphi_s$. Then $\psi_i \in \mathscr{M}_H(S)$ and, by Theorem 9(iii),

$$(\psi_i^G, \psi_j^G) = 1 + \delta_{ij}$$

for all $i, j \leqslant s-1$. We claim that there exist irreducible characters $\chi^{(1)}, \ldots, \chi^{(s)}$ of G and a sign $\varepsilon = \pm 1$ such that, for $i = 1, \ldots, s-1$,

$$\psi_i^G = \varepsilon(\chi^{(i)} - \chi^{(s)}). \qquad (1.2)$$

Since $\psi_i^G(1) = 0$, each ψ_i^G must be the difference of two irreducible characters. If $s = 2$ we have $\psi_1^G = \varepsilon(\chi^{(1)} - \chi^{(2)})$ with an arbitrary choice of labelling and sign. If $s \geqslant 3$, then we take $-\varepsilon\chi^{(s)}$ as the common constituent of ψ_1^G and ψ_2^G, and take $\chi^{(1)}$ and $\chi^{(2)}$ so that (1.2) holds for $i=1,2$. If $s \geqslant 4$, then we observe that, should $-\varepsilon\chi^{(s)}$ not be a constituent of ψ_i^G for some $i \geqslant 3$, then the inner products

$$(\psi_i^G, \psi_1^G) = (\psi_i^G, \psi_2^G) = 1$$

force

$$\psi_i^G = \varepsilon(\chi^{(1)} + \chi^{(2)}),$$

against the fact that $\psi_i^G(1) = 0$. It follows that no distinct ψ_i^G and ψ_j^G have any constituent in common other than $-\varepsilon\chi^{(s)}$, so (1.2) holds. We can now take differences to establish the first part of the theorem.

Finally, if χ is an irreducible character of G distinct from $\chi^{(1)}, \ldots, \chi^{(s)}$, then

$$(\chi|_H, (\varphi_i - \varphi_j))_H = (\chi, (\varphi_i - \varphi_j)^G)_G = (\chi, \varepsilon(\chi^{(i)} - \chi^{(j)}))_G = 0$$

by the Frobenius reciprocity theorem. Hence $(\chi|_H, \varphi_i) = (\chi|_H, \varphi_j)$.

Theorem 11. *Assume the hypothesis and conclusion of Theorem 10, with $s \geqslant 2$. Assume further that $\varphi_i(h) = 0$ for all i whenever h is an element of $H - S$ which is conjugate in G to an element of S. Then there exists a nonnegative integer α such that, for $i = 1, \ldots, s$,*

$$\varphi_i^G = \varepsilon\chi^{(i)} + \alpha \sum_{j=1}^{s} \chi^{(j)} + \Delta$$

where Δ is a character of G which is independent of i and does not contain any $\chi^{(j)}$ as a constituent.

Proof. From Theorem 10, it follows that, for each i,

$$\varphi_i^G - \varepsilon\chi^{(i)} = \varphi_s^G - \varepsilon\chi^{(s)}$$

and hence that there exist integers α_j such that, for all i,

$$\varphi_i^G = \varepsilon\chi^{(i)} + \sum_{j=1}^{s} \alpha_j\chi^{(j)} + \Delta$$

where either $\Delta = 0$ or Δ is a sum of characters of G not involving $\chi^{(1)}, \ldots, \chi^{(s)}$. Furthermore, since φ_i^G is actually a character, we have $\alpha_j \geqslant 0$ and $\alpha_j + \varepsilon \geqslant 0$ for all j. We claim that the α_js are all equal. We may compute that

$$\begin{aligned}\|\varphi_i^G\|^2 - \|\varphi_s^G\|^2 &= |G|^{-1} \sum_{g\in G} (\varphi_i^G(g)\overline{\varphi_i^G(g)} - \varphi_s^G(g)\overline{\varphi_s^G(g)}) \\ &= |G|^{-1}[G:H] \sum_{g\in S} (\varphi_i^G(g)\overline{\varphi_i^G(g)} - \varphi_s^G(g)\overline{\varphi_s^G(g)})\end{aligned}$$

since $\varphi_i^G(g) = \varphi_s^G(g)$ whenever g is not conjugate to an element of S, and deduce that

$$\|\varphi_i^G\|^2 - \|\varphi_s^G\|^2 = |H|^{-1} \sum_{g\in H} (\varphi_i^G(g)\overline{\varphi_i^G(g)} - \varphi_s^G(g)\overline{\varphi_s^G(g)}) = 0.$$

Thus

$$(\alpha_i + \varepsilon)^2 + \alpha_s^2 = \alpha_i^2 + (\alpha_s + \varepsilon)^2,$$

and hence, for all i,

$$\alpha_i = \alpha_s.$$

The characters $\chi^{(1)}, \ldots, \chi^{(s)}$ obtained in these theorems are called the *exceptional characters* associated with $\varphi_1, \ldots, \varphi_s$. If $s \geqslant 3$, we observe that there is a uniquely determined correspondence. These results should be compared with the calculations in Section 2.7 for the groups $\mathrm{SL}(2, 2^n)$, in particular formula (7.3), and also with Theorem 17. (See also Exercise 6.)

Under the hypothesis of Theorem 10, the linear independence of irreducible characters implies that S consists of at least s conjugacy classes in H. One the other hand, the rank of the submodule of $\mathscr{C}\hbar(H)$ spanned by the generalised characters $\varphi_i - \varphi_j$ for $1 \leqslant i, j \leqslant s$ is only $s - 1$. Hence, even if S is closed we have not considered the full module $\mathscr{M}_H(S)$. In the next section we shall study the isometry from $\mathscr{M}_H(S)$ to $\mathscr{C}\hbar(G)$ formally, and then in the remainder of the chapter give some applications. It turns out that only in very special situations can the full module $\mathscr{M}_H(S)$ be exploited, but the ideas behind Theorems 10 and 11 can be formalised and generalised. Separately, we observe that a consequence of the hypothesis of Theorem 10 is that all the characters φ_i have the same degree: in particular we shall wish to relax this condition. These questions will be studied in the next chapter.

Exercises

1. Let H be the dihedral group D_{2n}, n odd. Let S be the set of nonidentity elements of odd order in H. Let φ be the nonprincipal linear character and $\psi_1, \ldots, \psi_s$ ($s = \frac{1}{2}(n-1)$) the nonlinear characters. Show that the set $\{1 + \varphi - \psi_1, \psi_i - \psi_1\}_{i=2,\ldots,s}$ is a basis for $\mathscr{M}_H(S)$.
2. Let $H = NE$ be a Frobenius group with Frobenius kernel N. Let $S = N^\#$. Find a basis for $\mathscr{M}_H(S)$.
3. Let H be a generalised quaternion group and let S be the set of elements of order at least 4. Find a basis for $\mathscr{M}_H(S)$.
4. Let R be a principal ideal domain and let M be a finitely generated free R-module. Let $\{e_1, \ldots, e_n\}$ and $\{f_1, \ldots, f_n\}$ be R-bases for M and suppose that

$$e_i = \sum_{j=1}^{n} a_{ij} f_j$$

   for $a_{ij} \in R$. Prove that the $n \times n$ matrix $A = (a_{ij})$ is invertible over R.
5. Prove Proposition 8.

6. Let H be a dihedral subgroup of $G = \mathrm{SL}(2, 2^n)$ of order $2^n + 1$ ($n \geqslant 2$). Show that the nonidentity elements of H of odd order form a set of special classes. Let $\{\varphi_i\}$ be the nonlinear irreducible characters of H. By using the character table for $\mathrm{SL}(2, 2^n)$ determined in Section 2.7, decompose the induced characters $\{\varphi_i^G\}$ as sums of irreducible characters and compare the answer with Theorem 11.
   [Note that this example shows that the sign ε cannot be determined in general just from a knowledge of the subgroup H of that theorem.]
7. Let H be a group of exponent n and suppose that R is a principal ideal domain in an algebraic number field containing the nth roots of unity. Show that the conclusions of Lemma 5, Corollary 6 and Theorem 7 all hold in $\mathscr{C}\!h_R(H)$ without needing to assume that the normal subset C is closed.
   Show, by means of an example, that no analogue of Proposition 2 holds in $\mathscr{C}\!h_R(H)$.
   [Hint. Notice that the necessary analogue of Lemma 4 is trivially true.]

2 Suzuki's algorithm

Let H be a subgroup of a group G, and let S be a closed set of special classes in H consisting of t conjugacy classes. In the notation of the previous section, we shall take $R = \mathbb{Z}$ from now on. We shall examine the possible decompositions of θ^G as a linear combination of irreducible characters of G for $\theta \in \mathscr{M}_H(S)$.

Let $\varphi_1, \ldots, \varphi_n$ be the irreducible characters of H and let $\{\Phi_1, \ldots, \Phi_t\}$ be a $\mathbb{Z}$-basis for $\mathscr{M}_H(S)$. Then there exist integers a_{ik} such that

$$\Phi_i = \sum_{k=1}^{n} a_{ik}\varphi_k, \quad i = 1, \ldots, t,$$

so that the inner products are given by

$$(\Phi_i, \Phi_j) = \sum_{k=1}^{n} a_{ik}a_{jk}.$$

Writing A for the matrix (a_{ij}), it follows that the matrix whose (i, j)-entry is the inner product (Φ_i, Φ_j) is just AA^{T}, where A^{T} denotes the transpose of A.

Let $\chi_1, \ldots, \chi_r$ be the irreducible characters of G. Then there exist integers b_{il} such that

$$\Phi_i^G = \sum_{l=1}^{r} b_{il}\chi_l. \tag{2.1}$$

Thus

$$(\Phi_i^G, \Phi_j^G) = \sum_{l=1}^{r} b_{il}b_{jl}.$$

Writing $B = (b_{ij})$, the isometry from $\mathscr{M}_H(S)$ to $\mathscr{C}\!h(G)$ given by character induction yields the matrix equation

$$AA^{\mathrm{T}} = BB^{\mathrm{T}}. \tag{2.2}$$

Suppose now that $\chi_1, \ldots, \chi_r$ are ordered so that

$$B = (B_0\ 0)$$

where the columns of B_0 are nonzero. Since the inner products of rows of B are known from (2.2) and the entries in B are integers, each row of B has a bounded number of nonzero entries, themselves bounded, so that B_0 has a bounded number of nonzero columns and bounded, integral entries. Formally this provides an algorithm, but we have in particular shown the following.

Proposition 12. *Only finitely many possibilities for B_0 exist.*

We now come to the main result of this section, that the matrix B determines the values of the characters of G on S.

Theorem 13. *For each possible matrix B given by* (2.1) *and* (2.2), *there exists an integral matrix $X = (x_{ij})$ such that $B = AX$. The values $\chi_i(s)$, for $s \in S$, are uniquely determined by B and, for* any *matrix X for which $B = AX$, those values are given by*

$$\chi_i(s) = \sum_{k=1}^{n} x_{ki}\varphi_k(s).$$

Proof. Since $\mathscr{M}_H(S)$ is a direct summand of $\mathscr{C}\hbar(H)$ of rank t, there exist unimodular integral matrices P and Q such that

$$A = P(I_t\, 0)Q$$

where I_t is the identity $t \times t$ matrix. Then $B = AX$ where

$$X = Q^{-1}\begin{pmatrix} P^{-1}B \\ 0 \end{pmatrix}.$$

Now let $X = (x_{ij})$ be any matrix such that $B = AX$. For any fixed i, $1 \leqslant i \leqslant r$, put

$$\psi = \chi_i|_H - \sum_{k=1}^{n} x_{ki}\varphi_k.$$

Then we need only show that $\psi \in \mathscr{M}_H(\hat{S})$ or equivalently, since S is closed, that $\psi \in \hat{\mathscr{M}}_H(S)$. Thus it is sufficient to show that $(\psi, \Phi_j)_H = 0$ for $j = 1, \ldots, t$. Now, application of the Frobenius reciprocity theorem shows that

$$(\psi, \Phi_j)_H = \left(\left(\chi_i|_H - \sum_{k=1}^{n} x_{ki}\varphi_k\right), \Phi_j\right)_H$$

$$= (\chi_i|_H, \Phi_j)_H - \sum_{k=1}^{n} x_{ki}(\varphi_k, \Phi_j)_H$$
$$= (\chi_i, \Phi_j^G)_G - \sum_{k=1}^{n} x_{ki}a_{jk}$$
$$= b_{ji} - \sum_{k=1}^{n} a_{jk}x_{ki}$$
$$= 0$$

since $B = AX$.

When relating this result to our initial goal, that of showing that in suitable situations a portion of the character table of H lifts into the character table of G, we note that in the presence of a closed set of special classes there is a natural one-to-one map from the set of H-conjugacy classes contained in S into the set of G-conjugacy classes. An 'ideal' situation, where the relevant portion of the character table just lifts, would be given by $B_0 = A_0$, where A_0 consists of the nonzero columns of A, so that X could be taken as, essentially, an identity matrix. However, if J is any diagonal matrix with diagonal entries ± 1, then BJ will be a solution of the matrix equation (2.2) whenever B is. This corresponds to replacing an irreducible character of G by its negative, something that we cannot distinguish when working with isometries alone. Any such situation can be regarded as a *natural induction.* However, no claim can be made in general about the uniqueness of the matrix B, even up to signs and interchanges of columns. Even when such uniqueness can be established, use of the isometry alone is normally insufficient and it is necessary to apply the Frobenius reciprocity formula to inner products involving the principal character in order to provide additional relations for this purpose.

3 The Brauer–Suzuki theorem

Our first application of these methods will require only Theorem 9. The group given by the presentation

$$\langle x, y \mid x^{2^n} = 1, y^2 = x^{2^{n-1}}, y^{-1}xy = x^{-1} \rangle \qquad (3.1)$$

is a nonabelian group of order 2^{n+1} if $n \geqslant 2$. If $n = 2$, it is called a *quaternion* group: if $n \geqslant 3$, it is a *generalised* quaternion group. In either case, it is the unique noncyclic group of its order containing exactly one involution. This group and its representations were described in the exercises in Section 1.1. Brauer and Suzuki proved that a finite group having such a group as Sylow 2-subgroup cannot be simple. Since Burnside's transfer theorem (Corollary 1.48) will show that a group having a nontrivial cyclic Sylow 2-subgroup has a normal 2-complement, this implies that a finite simple group (of even order) must contain a subgroup isomorphic to $Z_2 \times Z_2$.

We shall give a proof of this result in the case of a generalised quaternion group in this section. For quaternion groups of order 8, the original proof given by Brauer and Suzuki (1959) depended on the use of modular character theory, as does the refinement by Brauer (1964; II). In Section 6.6, we shall give the proof outlined by Suzuki (1962a) since it provides an interesting application of the particular ideas that we shall be developing in that chapter; also we shall be able to reduce that situation to precisely the same calculation that occurs in the generalised quaternion group case. (Specifically, we shall reach the point (3.2).) Later, Glauberman (1974) was able to give a proof in this case using ordinary character theory. However, his method is to follow the path of the modular theory proof and involves a detailed discussion of character values and an ingenious application of the Frobenius–Schur indicator rather than the method of exceptional characters. In the exercise at the end of this section, we will give an example where exceptional character theory as we have developed it in this chapter can be used for this case.

Theorem 14. *Let G be a finite group having a Sylow 2-subgroup T which is a generalised quaternion group of order at least 16. Assume that G has no nonidentity normal subgroup of odd order. Then $|Z(G)| = 2$.*

Proof. Let T be generated by elements x, y satisfying the presentation (3.1). Then it is easily verified that every element of $T - \langle x \rangle$ has order 4 so that $\langle x \rangle$ is the unique cyclic subgroup of index 2 in T and hence is characteristic in T. Let $u = x^{2^{n-2}}$ and $z = x^{2^{n-1}}$, and put $\langle x \rangle = X$, $\langle u \rangle = U$ and $\langle z \rangle = Z$. Then U is the unique subgroup of X of order 4, and $Z = Z(T)$. Let $K = C_G(u)$ and $H = N_G(U)$. Then X is a Sylow 2-subgroup of K so that K has a normal 2-complement N. Thus $K = XN$ and $H = TN$.

Let $L = ZN$. We claim that $K - L$ forms a closed set of special classes in H. Certainly it is a normal subset, and it is closed since $K - L$ consists precisely of those elements of K whose order is divisible by 4. Now $K \subseteq H$, and any element of order 4 in $K - L$ is conjugate in H to u. Thus, if $a, b \in K - L$ and $b = a^g$ for some $g \in G$, by replacing a and b by suitable powers we may assume that a and b have order 4, whence there exist $g_1, g_2 \in H$ with $a = u^{g_1}$ and $b = u^{g_2}$ so that $g_1 g g_2^{-1} \in K$ and $g \in H$.

Let φ be the nonprincipal character of H with kernel K, and let ψ be the character of H with kernel L afforded by the representation

$$x \to \begin{pmatrix} \omega & 0 \\ 0 & \omega^{-1} \end{pmatrix}, \quad y \to \begin{pmatrix} 0 & -1 \\ 1 & 0 \end{pmatrix},$$

where ω is a primitive 2^{n-1}st root of unity. (See Exercise 10 of Section 1.1.)

Since $n \geqslant 3$, this representation is irreducible. (Note that T/Z is a dihedral group.) Then we see that the generalised character θ given by

$$\theta = 1_H + \varphi - \psi \tag{3.2}$$

lies in $\mathcal{M}_H(K - L)$. Since $\|\theta\|_H^2 = 3$, it follows from Theorem 9 that $\|\theta^G\|_G^2 = 3$ and that $\theta^G(1) = 0$. Also, by the Frobenius reciprocity formula,

$$(\theta^G, 1_G)_G = (\theta, 1_H)_H = 1.$$

Thus there exist irreducible characters Φ, Ψ of G such that

$$\theta^G = 1_G + \Phi - \Psi. \tag{3.3}$$

If two involutions of G were to have a product of even order, then G would contain a subgroup isomorphic to $Z_2 \times Z_2$ by Proposition 2.52, which is impossible. Thus G contains no strongly real elements of even order. Define a function $\xi \in \mathscr{C}(G, \mathbb{C})$ by

$$\xi(g) = \sum_{\chi} \frac{\chi(z)^2}{\chi(1)} \chi(g) \tag{3.4}$$

where the summation is taken over all irreducible characters of G. If $\xi(g) \neq 0$, then g is strongly real, and so ξ vanishes on all elements of G of even order. Now θ^G vanishes on all elements of odd order: thus $(\xi, \theta^G) = 0$. From (3.3) and (3.4), we obtain the equations

$$1 + \frac{\Phi(z)^2}{\Phi(1)} - \frac{\Psi(z)^2}{\Psi(1)} = 0,$$

$$\Psi(1) = 1 + \Phi(1) \quad \text{and} \quad \Psi(z) = 1 + \Phi(z),$$

the last since $z \notin K - L$. From these equations it follows that $\Phi(z) = \Phi(1)$ and that $\Psi(z) = \Psi(1)$, as in (9.4) of Section 2.9.

In particular, G cannot be simple but, furthermore,

$$z \in \ker \Phi \cap \ker \Psi.$$

On the other hand,

$$\theta^G(u) = \theta(u) = 2:$$

thus $u \notin \ker \Phi \cap \ker \Psi$. Now any normal subgroup of G of order divisible by 4 contains u. So either $T \cap \ker \Phi = Z$ or $T \cap \ker \Psi = Z$ (or both). In the first case, $\ker \Phi$ has twice odd order and so has a normal 2-complement. Since G has no normal subgroup of odd order, this implies that $\ker \Phi = Z$, so that Z is normal in G. The same argument applies in the second case. Now $Z \subseteq Z(G)$, but it is clear that $Z(G)$ cannot be larger than Z, and the proof is complete.

Exercise

1. Let G be a finite group with $O(G) = 1$, and suppose that the Sylow 2-subgroups are quaternion of order 8. Let z be an involution, put

$H = C_G(z)$ and suppose that $H/O(H) \cong \mathrm{SL}(2, 3)$ or $\mathrm{SL}(2, 5)$. Show that the elements of even order in H form a set of special classes in H. Show that $\mathrm{SL}(2, 3)$ and $\mathrm{SL}(2, 5)$ have irreducible characters χ and ψ such that, as characters of H, $1_G + \chi - \psi$ vanishes off the special classes.

Show that G has no strongly real elements of even order and deduce that G cannot be simple.

4 Strongly self-centralising subgroups

We begin with a definition very closely related to that of a set of special classes.

Definition. A subset A of a group G is a *trivial intersection subset* (or *T.I.-set*) in G if the following conditions hold:

(i) $A \subseteq N_G(A)$;
(ii) $A \cap A^g \subseteq \{1\}$ whenever $g \in G - N_G(A)$.

Thus, if A is a T.I.-set in G, then $A - \{1\}$ forms a set of special classes in $N_G(A)$ by Proposition 8. It is a trivial exercise to verify that the subgroup A of the following definition is a T.I.-set.

Definition. A nonidentity abelian subgroup A of a group G is said to be *strongly self-centralising* if $C_G(a) = A$ for every nonidentity element a of A.

We remark that, in a group G in which the centraliser of every nonidentity element of odd order is abelian, the maximal abelian subgroups of odd order will be strongly self-centralising. This is the situation which will occur in our discussion of the so-called CA-groups of odd order in the next section: also we have seen this condition hold in the groups $\mathrm{SL}(2, 2^n)$ and we will give a characterisation based on this property at the end of the section.

Because we shall consider generalisations and related questions in the next chapter, we shall derive the basic group theoretical consequence of this definition in a broader setting.

Proposition 15. *Let G be a group and let H be a subgroup of G such that $C_G(h) \subseteq H$ for all $h \in H^\#$. Then H is a Hall subgroup of G. If $N = N_G(H) \neq H$, then N is a Frobenius group with Frobenius kernel H.*

Proof. Let p be a prime divisor of $|H|$, and let P_1 be a Sylow p-subgroup of H. If P is a Sylow p-subgroup of G containing P_1, then $Z(P) \subseteq C(P_1) \subseteq H$, whence $P \subseteq C(Z(P)) \subseteq H$. So H is a Hall subgroup of G. The final statement follows from Proposition 2.42.

Suppose now that A is a strongly self-centralising subgroup of a group G. If $N_G(A) = A$, then Burnside's transfer theorem will show that A has a normal complement in G, and then, if $G \neq A$, the group G will itself be a Frobenius group with Frobenius complement A. (Indeed, a transfer argument will establish a similar conclusion if $H = N_G(H)$ in Proposition 15.) Assume, then, that $N_G(A) \neq A$. Let $N = N_G(A) = AE$ where E is a Frobenius complement, and put $e = |E|$ and $t = (|A| - 1)/e$. By Theorem 2.43, N has precisely t irreducible characters $\varphi_1, \ldots, \varphi_t$, each of degree e, which do not contain A in their kernels, and each vanishes on $N - A$. If $t \geqslant 2$, the hypothesis of Theorem 10 is satisfied, and G possesses t exceptional characters $\chi^{(1)}, \ldots, \chi^{(t)}$ associated with $\varphi_1, \ldots, \varphi_t$, and their values on A will be given by the following.

Theorem 16. *In the situation described above with $e \geqslant 2$ and $t \geqslant 2$, there exist a sign $\varepsilon = \pm 1$ and a rational integer c such that, for each element $a \in A^{\#}$,*

$$\chi^{(i)}(a) = \varepsilon\varphi_i(a) + c, \quad i = 1, \ldots, t.$$

All other characters are constant and rational-valued on $A^{\#}$. For any element g of G which is not conjugate to an element of $A^{\#}$, the exceptional characters $\chi^{(1)}, \ldots, \chi^{(t)}$ take the same rational value: in particular, they have the same degree.

Proof. It is an immediate consequence of Theorems 9(i) and 10 that there exist characters $\chi^{(1)}, \ldots, \chi^{(t)}$ of G, a sign $\varepsilon = \pm 1$, and constants c_a, possibly depending on a, such that

$$\chi^{(i)}(a) = \varepsilon\varphi_i(a) + c_a$$

for $a \in A^{\#}$ and $i = 1, \ldots, t$, where $c_a = \chi^{(1)}(a) - \varepsilon\varphi_1(a)$. Let $\lambda_1, \ldots, \lambda_f$ be the distinct irreducible characters of N which have A in their kernels, and write

$$\chi^{(1)}|_N = \sum_{j=1}^{t} b_j\varphi_j + \sum_{k=1}^{f} d_k\lambda_k$$

for nonnegative integers $\{b_j\}$, $\{d_k\}$. Then, by Theorem 11, there is a nonnegative integer α such that

$$b_j = (\chi^{(1)}|_N, \varphi_j)_N = (\chi^{(1)}, \varphi_j^G)_G = \varepsilon\delta_{1j} + \alpha,$$

and thus, for $a \in A^{\#}$,

$$c_a = \chi^{(1)}(a) - \varepsilon\varphi_1(a) = \alpha \sum_{j=1}^{t} \varphi_j(a) + \sum_{k=1}^{f} d_k\lambda_k(1).$$

Applying the orthogonality relations to N, we obtain the equations

$$\sum_{k=1}^{f} \lambda_k(1)^2 + e \sum_{j=1}^{t} \varphi_j(a) = 0$$

and

$$\sum_{k=1}^{f} \lambda_k(1)^2 = e,$$

so that c_a is a rational integer, independent of a.

Now let χ be any irreducible character of G distinct from $\chi^{(1)}, \ldots, \chi^{(t)}$. Then, by Theorem 10 and the Frobenius reciprocity theorem,

$$(\chi|_N, (\varphi_i - \varphi_j))_N = (\chi, \varepsilon(\chi^{(i)} - \chi^{(j)}))_G = 0.$$

Hence, for some integer k and an integral linear combination λ of the characters $\lambda_1, \ldots, \lambda_f$,

$$\chi|_N = k \sum_{i=1}^{t} \varphi_i + \lambda$$

so that, for $a \in A^{\#}$,

$$\chi(a) = -k + \lambda(1).$$

Finally, we must consider the values of the exceptional characters outside $A^{\#}$. It follows from Theorem 10 that they all take the same values. Suppose that some value were irrational, say $\chi^{(i)}(g)$. Since g has order coprime to $|A|$, there would be a Galois automorphism θ of some finite normal extension of $\mathbb{Q}$ fixing all $|A|$th roots of 1 but moving $\chi^{(i)}(g)$. Now each exceptional character is characterised by its (nonconstant) values on $A^{\#}$, and hence $(\chi^{(i)})^{\theta} = \chi^{(i)}$, which is impossible.

In terms of our original goal, to lift part of the character table of N to the character table of G, we would like the constant c which occurs in Theorem 16 to be zero. In the next chapter, using far deeper methods we shall prove this, provided that $t \geqslant 3$. Indeed, we shall also obtain an analogue of Theorem 16 starting with a nonabelian group H which is a T.I.-set and which satisfies the hypothesis of Proposition 15. However, in the case that e is very small, one can establish that $c = 0$ without the limitation on t by the methods of Section 2 alone, even allowing for the possibility that $t = 1$.

Again, suppose that A is a strongly self-centralising subgroup of a group G, let $N = N_G(A)$, and let e and t be defined as before. Then $A^{\#}$ consists of t conjugacy classes of N so that the module $\mathscr{M}_N(A^{\#})$, as defined in Section 1, has rank t. If $\varphi_1, \ldots, \varphi_t$ are the characters of N as defined above, then the generalised characters $\{\varphi_i - \varphi_j\}$ span a submodule of rank only $t-1$. However, if π denotes the character of N afforded by the regular representation of N/A so that

$$\pi(g) = \begin{cases} 0 & \text{if } g \in N - A, \\ e & \text{if } g \in A, \end{cases}$$

then the generalised characters $\pi - \varphi_1, \varphi_1 - \varphi_2, \ldots, \varphi_1 - \varphi_t$ form a basis for $\mathcal{M}_N(A^\#)$. To see this, we need only observe that, in order that a generalised character θ of N should vanish on $N - A$, it must be of the form

$$\theta = c\pi + \sum_{j=1}^{t} d_j\varphi_j,$$

and then, if $\theta(1) = 0$ also, we obtain the equation

$$c + \sum_{j=1}^{t} d_j = 0,$$

whence

$$\theta = c(\pi - \varphi_1) - \sum_{j=2}^{t} d_j(\varphi_1 - \varphi_j).$$

Let $\lambda_1 = 1_N, \lambda_2, \ldots, \lambda_f$ be the irreducible characters of N which have A in their kernels. Then $\pi = \sum_i \lambda_i(1)\lambda_i$. Thus to have a natural induction in the sense of Section 2, we need to find irreducible characters $\Lambda_1 = 1_G$, $\Lambda_2, \ldots, \Lambda_f$ of G, distinct from the exceptional characters $\chi^{(1)}, \ldots, \chi^{(t)}$ associated with $\varphi_1, \ldots, \varphi_t$ (assuming $t \geqslant 2$) and signs $\varepsilon, \varepsilon_2, \ldots, \varepsilon_f = \pm 1$ such that

$$(\pi - \varphi_1)^G = 1_G + \sum_{i=2}^{f} \varepsilon_i\lambda_i(1)\Lambda_i - \varepsilon\chi^{(1)} \tag{4.1}$$

and

$$(\varphi_1 - \varphi_j)^G = \varepsilon(\chi^{(1)} - \chi^{(j)}), \quad j = 2, \ldots, t. \tag{4.2}$$

The equations (4.2) come from Theorem 10; the multiplicities of 1_G and $\chi^{(1)}$ in equation (4.1) are forced by applying the Frobenius reciprocity formula to the inner product $((\pi - \varphi_1)^G, 1_G)_G$ and by considering the underlying isometry which yields the inner product

$$((\pi - \varphi_1)^G, (\varphi_1 - \varphi_2)^G)_G = (\pi - \varphi_1, \varphi_1 - \varphi_2)_N = -1.$$

In the situation where the equations (4.1) and (4.2) do hold, Theorem 13 yields the following character values on $A^\#$:

$$\Lambda_i(a) = \varepsilon_i, \qquad i = 2, \ldots, f,$$
$$\chi^{(j)}(a) = \varepsilon\varphi_j(a), \quad j = 1, \ldots, t.$$

In particular, the constant c of Theorem 16 is zero.

The problem, then, is to obtain the decomposition (4.1). Unfortunately this is impossible in general simply from the information immediately available, namely that

$$\|(\pi - \varphi_1)^G\|^2 = e + 1,$$
$$(\pi - \varphi_1)^G(1) = 0,$$
$$((\pi - \varphi_1)^G, (\varphi_1 - \varphi_j)^G) = -1, \quad j = 2, \ldots, t,$$

and

$$((\pi - \varphi_1)^G, 1_G) = 1.$$

The obstruction lies in the fact that we cannot control the multiplicities of the individual constituents of $(\pi - \varphi_1)^G$ unless $e \leqslant 4$, in which case they must be ± 1. As an illustration, we shall consider explicitly the case $e = 2$. However, it should be mentioned that, in the special case where some Sylow subgroup of A is cyclic, deep results of Dade in modular representation theory (on blocks with a cyclic defect group) enable one to establish the decomposition (4.1) for $(\pi - \varphi_1)^G$ without limitation on e. (See, for example, Alperin [A].)

Theorem 17. *Let A be a strongly self-centralising subgroup of a group G, let $N = N_G(A)$, and assume that $[N:A] = 2$. Let $t = \frac{1}{2}(|A| - 1)$ and let $\varphi_1, \ldots, \varphi_t$ be the irreducible characters of N of degree 2. Then there exist characters $\chi, \chi^{(1)}, \ldots, \chi^{(t)}$ of G and a sign $\varepsilon = \pm 1$ such that the following hold.*

(i) $\chi(a) = \varepsilon$ *if* $a \in A^{\#}$.
(ii) χ *is rational-valued.*
(iii) $\chi^{(i)}(a) = \varepsilon\varphi_i(a)$ *if* $a \in A^{\#}$, *for* $i = 1, \ldots, t$.
(iv) *Every other nonprincipal irreducible character of G vanishes on $A^{\#}$ and has degree divisible by $|A|$.*
(v) *For any element g of G, not conjugate to an element of $A^{\#}$,*

$$\chi^{(i)}(g) = \chi(g) + \varepsilon, \quad i = 1, \ldots, t.$$

(vi) $\chi(1) \equiv \varepsilon \pmod{|A|}$.

Proof. Let λ be the nonprincipal linear character of N and put

$$\theta = 1_G + \lambda - \varphi_1.$$

Then $\|\theta^G\|^2 = 3$, $\theta^G(1) = 0$ and $(\theta^G, 1_G) = (\theta, 1_N) = 1$. Thus there exist nonprincipal irreducible characters η and ζ of G such that

$$\theta^G = 1_G + \eta - \zeta. \tag{4.3}$$

We first establish (i), (iii) and (v).

If $|A| = 3$, we simply put $\chi = \eta$ and $\chi^{(1)} = \zeta$: then $\varepsilon = 1$ is forced by Theorem 13 and the fact that $\theta^G(a) = 3$ for $a \in A^{\#}$.

If $|A| = 5$, we have

$$(\varphi_1 - \varphi_2)^G = \varepsilon(\chi^{(1)} - \chi^{(2)}) \tag{4.4}$$

and a choice for the sign ε. Since

$$(\theta^G, (\varphi_1 - \varphi_2)^G) = -1,$$

we choose $\chi^{(1)}$ and $\chi^{(2)}$ in (4.4) so that $-\varepsilon\chi^{(1)}$ is a constituent of θ^G: then we take χ as that irreducible character of G for which

$$\theta^G = 1_G + \varepsilon\chi - \varepsilon\chi^{(1)}, \tag{4.5}$$

noting that the sign is forced by the fact that $\theta^G(1)=0$. Again, Theorem 13 applies to give (i) and (iii), and then (v) follows from (4.5).

If $|A|\geqslant 7$, then there is a uniquely determined sign $\varepsilon=\pm 1$ such that

$$(\varphi_1-\varphi_j)^G=\varepsilon(\chi^{(1)}-\chi^{(j)}),\quad j=2,\ldots,t,$$

by Theorem 10. Since

$$(\theta^G,(\varphi_1-\varphi_j)^G)=-1,\quad j=2,\ldots,t,$$

$-\varepsilon\chi(1)$ must be a constituent of θ^G if there are at least three such inner products, namely if $t\geqslant 4$, and the proof of (i), (iii) and (v) can proceed as before. If $t=3$, the inner products yield an additional possibility, namely that the common constituent of θ^G and $(\varphi_1-\varphi_2)^G$ is $\chi^{(2)}$ and that the common constituent of θ^G and $(\varphi_1-\varphi_3)^G$ is $\chi^{(3)}$. But then

$$\theta^G=1_G+\varepsilon\chi^{(2)}+\varepsilon\chi^{(3)},$$

which is impossible since $\theta^G(1)=0$. So again we may proceed as before.

Now, for $a\in A^{\#}$,

$$1+|\chi(a)|^2+\sum_{i=1}^{t}|\chi^{(i)}(a)|^2=2+\sum_{i=1}^{t}|\varphi_i(a)|^2=|A|=|C_G(a)|;$$

if η is any other nonprincipal character of G, this forces $\eta(a)=0$ and then

$$(\eta|_A,1_A)_A=|A|^{-1}\eta(1)\in\mathbb{Z},$$

while

$$(\chi|_A,1_A)_A=|A|^{-1}(\chi(1)+\varepsilon(|A|-1))\in\mathbb{Z},$$

from which we obtain (iv) and (vi). Finally, to see (ii), we observe that, for any i, there exists $a\in A^{\#}$ such that $\chi(a)\neq\chi^{(i)}(a)$. Thus, by (iv), the character χ can have no algebraic conjugate other than itself, and so is rational-valued.

We remark that the special arguments needed to handle the cases $|A|=3$ and 5 are not untypical; we will frequently see the distinction between 'small' and 'generic' cases.

Corollary 18. *Assume the hypothesis of Theorem* 17 *and suppose also that G is simple. Then G contains exactly one conjugacy class of involutions.*

Proof. If $a\in A^{\#}$ and $a^g\in A$, then $g\in N$. Also, since N has twice odd order, all involutions in N are conjugate. Suppose, then, that G contains more than one conjugacy class of involutions. Let $\mathscr{C}$ be a class of involutions in G disjoint from N. Then no involution in $\mathscr{C}$ can invert a and hence, in the notation of Section 2.9, $\mathscr{A}(\mathscr{C},\mathscr{C};a)=\varnothing$. Computing the corresponding class algebra as in Example 2 of that section with the character values given by Theorem 17, we obtain the equation

$$1+\frac{\varepsilon\chi(\tau)^2}{\chi(1)}-\frac{\varepsilon(\chi(\tau)+\varepsilon)^2}{\chi(1)+\varepsilon}=0$$

where $\tau\in\mathscr{C}$. From this, as in (9.4) of Section 2.9, we may deduce that $\chi(\tau)=\chi(1)$, which is impossible if G is simple.

We remark that, as a consequence of a theorem of Smith and Tyrer (1973), the hypothesis of Theorem 17 together with simplicity implies that A must be cyclic. Their theorem requires results from block theory, although in the special case that A is a p-group, a proof can be given using results of this chapter. Under the assumption that A is noncyclic, results from cohomology show that G will have a perfect central extension $\hat{G}$ by a cyclic subgroup of order p in which the extension $\hat{A}$ of A itself does not split. The elements of $\hat{A}-Z(\hat{G})$ form a set of special classes in $\hat{N}$ and, ultimately, a contradiction can be obtained by the class algebra constant method, observing that no involution can invert $Z(\hat{G})$.

We complete this section with an application of Theorem 17 which gives a further characterisation of the simple groups $\mathrm{SL}(2,2^n)$, due independently to Harada (1967) and Stewart (1967); we shall follow the latter's approach.

Theorem 19. *Let G be a finite simple group which possesses two nonconjugate strongly self-centralising subgroups A and B with $[N(A){:}A]=[N(B){:}B]=2$. Then G is isomorphic to $\mathrm{SL}(2,2^n)$ for some $n\geqslant 2$.*

Proof. Without loss, we may take $|A|>|B|$ and adopt the notation of Theorem 17 for A. Then none of the characters $\chi^{(i)}$ is rational-valued on $A^\#$, nor do all such characters take the same values on any element of $A^\#$.

Let $s=\frac{1}{2}(|B|-1)$ and let $\zeta,\zeta^{(1)},\ldots,\zeta^{(s)}$ have the corresponding meaning for B as do $\chi,\chi^{(1)},\ldots,\chi^{(t)}$ for A. By the remarks above, no character $\chi^{(i)}$ can be any ζ or $\zeta^{(j)}$. Hence, since the columns of character values at elements of $A^\#$ and $B^\#$ are orthogonal, it follows that $\chi=\zeta$ and that, giving ε_A and ε_B their obvious meanings, $\varepsilon_A=-\varepsilon_B$. (If $|B|=3$, then ζ and $\zeta^{(1)}$ are distinguished only by this fact.)

Let $\varepsilon=\varepsilon_B$ and $\chi(1)=x$. By Theorem 17(vi), $x\equiv-\varepsilon\ (\mathrm{mod}\,|A|)$ and $x\equiv\varepsilon\ (\mathrm{mod}\,|B|)$ and, in particular, $x\neq 1$. Let $\mathscr{C}$ be the unique class of involutions in G. Then $|\mathscr{A}(\mathscr{C},\mathscr{C};a)|=|A|$ and $|\mathscr{A}(\mathscr{C},\mathscr{C};b)|=|B|$ for $a\in A^\#$ and $b\in B^\#$ where $\mathscr{A}(\mathscr{C},\mathscr{C};a)$ has the same meaning as in Section 2.9. Computing with the standard character theoretic formulae for each of the corresponding class algebra constants, we can deduce that

$$(x-\varepsilon)|A|=(x+\varepsilon)|B|.$$

Table 3.1

	1	$a \in A^{\#}$	$b \in B^{\#}$	τ	g
1	1	1	1	1	1
χ	x	-1	1	u	v
$\chi^{(i)}$	$x-1$	$-\varphi_i(a)$	0	$u-1$	$v-1$
$\zeta^{(j)}$	$x+1$	0	$\psi_j(b)$	$u+1$	$v+1$

Since $|A| > |B|$, it follows that $\varepsilon = 1$, and then that

$$|A| = x + 1 \quad \text{and} \quad |B| = x - 1. \tag{4.6}$$

This gives rise to the fragment of character table shown in Table 3.1, where the nonlinear characters of $N(B)$ are $\{\psi_j\}$, τ is an involution, and g is any element of G not conjugate to an element of $A^{\#} \cup B^{\#}$. Also, we put $u = \chi(\tau)$ and $v = \chi(g)$.

By Theorem 17(iv), every other irreducible character of G has degree divisible by $|A| \cdot |B|$; hence the product character $\chi\chi^{(1)}$ must be a nonnegative linear combination of those characters displayed in this table only. We could, in fact, determine it from the values on $A^{\#} \cup B^{\#}$ (exercise!); however, notice that the value on an element g gives a quadratic expression of the form

$$v(v-1) = \alpha v + \beta$$

where α and β are independent of the choice of g. Now one solution is $v = x$; since $x \neq 1$, χ vanishes on some element of G by a result of Burnside (Exercise 26 of Section 2.1), so that the other root is $v = 0$. But now $\chi(g) = x$ only if $g = 1$ since G is simple; hence $\chi(g) = 0$ whenever g is not conjugate to an element of $A \cup B$. In particular, $\chi(\tau) = 0$.

Now the inner product $(\chi, \chi) = 1$, together with (4.6), yields the formula

$$|G| = x(x-1)(x+1),$$

whence summing the squares of degrees shows that no further characters exist. Since the number of conjugacy classes is equal to the number of irreducible characters, we see that every element of G is conjugate to an element of $N(A) \cup N(B)$; in particular, it follows that a Sylow 2-subgroup of G is elementary abelian of order x and is precisely the centraliser of each of its involutions. The characterisation is now given by Theorem 2.56.

We shall give a further explicit application of Theorem 17 in our discussion of part of the classification of Zassenhaus groups in Section 4.5.

Exercises

1. Let G be a simple group in which the normaliser of a Sylow 3-subgroup P is a Frobenius group of order 36. Suppose that P is a T.I.-set in G.

Show that there are precisely six irreducible characters of G which do not vanish on $P^{\#}$ and determine their values on $P^{\#}$ and the relations between their degrees.

[Compare this with the calculation in Example 3 in Section 2.9.]

2. (Camina and Collins (1974).) Let $N = HT$ be a group of order 108, where H is a normal elementary abelian subgroup of order 27 and T is elementary abelian of order 4. Suppose that $|C_H(\tau)| = 3$ for each $\tau \in T^{\#}$. show that N has 15 conjugacy classes, of which 11 contain elements of order divisible by 3.

   Suppose that N is a subgroup of a group G and that $C_G(h) \subseteq N$ for any element $h \in H^{\#}$. Let S denote the set of elements of N of order divisible by 3. Show that S is a set of special classes, and find a basis for $\mathscr{M}_N(S)$. Show also that a natural induction occurs.

   Use Proposition 2.53(ii) (as in Example 3 of Section 2.9) to determine character degrees, and deduce that $G = N$.

3. Determine the character $\chi\chi^{(1)}$ in the proof of Theorem 19 as a linear combination of the given irreducible characters.
4. Show that the proof of Theorem 19 can be completed by following the argument of the proof of Theorem 2.44, rather than depending upon the fuller analysis required for the proof of Theorem 2.56.

5 CA-groups of odd order

A finite group is said to be a *CA-group* if the centraliser of every nonidentity element is abelian. In this section, we shall give an account of one of the early applications of the method of exceptional characters, due to Suzuki (1957).

Theorem 20. *There do not exist nonabelian simple CA-groups of odd order.*

This result was the forerunner of the ideas involved in the proof of the solubility of groups of odd order. As a next step, Feit, Hall and Thompson (1960) considered CN-groups of odd order; a *CN-group* has the centralisers of all nonidentity elements nilpotent. In this latter case, the group theoretical reductions are far more complex and their character theoretical arguments were very different from Suzuki's: however, it is now possible to give a character theoretical argument similar to Suzuki's, and we shall discuss this in Section 4 of the next chapter, while in this section we shall indicate the points at which the additional difficulties will arise. As we shall see, the basic group theoretic goal is straightforward; character theory then yields a delicate numerical contradiction which has no obvious group theoretic interpretation.

The first stage of the proof is a reduction to the point where the results in Theorem 16 apply. Assume then, for the remainder of the section, that G is a nonabelian simple CA-group of odd order, and let A be a maximal abelian subgroup of G. Clearly $A = C_G(a)$ for each $a \in A^\#$ and hence A is strongly self-centralising. By Proposition 15, A is a Hall subgroup of G and, by Burnside's transfer theorem, $A \neq N_G(A)$. By considering the maximal abelian subgroups of G, it is easy to see that G must have the following properties.

Proposition 21. *The prime divisors of $|G|$ fall into disjoint subsets $\pi_1, \ldots, \pi_s$ such that, for each i,*

(i) *there is an abelian Hall π_i-subgroup H_i of order h_i which is strongly self-centralising, and*
(ii) *if $N_i = N_G(H_i)$, then N_i is a Frobenius group of order $h_i e_i$ with $e_i \geqslant 3$ and H_i the Frobenius kernel.*

Before proceeding we remark that, for CN-groups, the group theoretic goal will be essentially the same, namely to show that H_i is a T.I.-set: however, Burnside's theorem is not applicable and, especially in the case where a nonabelian Sylow p-subgroup P forms a maximal nilpotent subgroup, there is no immediate reason why an element of $Z(P)$ might not be conjugate to an element of $P - Z(P)$. (The curious reader should try to show this to be impossible by a completely bare hands approach before reading the next chapter!)

Let G be a simple CA-group and adopt the notation of Proposition 21. Then every element of G is conjugate to an element in one of the subgroups H_i. Put $t_i = (h_i - 1)/e_i$: then $t_i \geqslant 2$ since $(h_i - 1)$ is even. So we may apply Theorem 16. For each i, let $\{\varphi_{i1}, \ldots, \varphi_{it_i}\}$ be the characters of N_i which do not have H_i in their kernels, and let $\chi_{i1}, \ldots, \chi_{it_i}$ be the corresponding exceptional characters of G. Since G has odd order, the characters $\varphi_{i1}, \ldots, \varphi_{it_i}$ are nonreal and hence each exceptional character χ_{ij} is nonreal on some element of $H_i^\#$ but is real (and rational) on every element of H_k for $k \neq i$. Thus χ_{ij} is exceptional for the subgroup H_i but for no other. Since the number of characters $\{\chi_{ij}\}$ is equal to the number of nonidentity conjugacy classes in G, together with 1_G we have all the irreducible characters of G and, as a consequence, we may organise the character table of G as in Table 3.2. Here, the entries in a block A_{ii} are of the form

$$\chi_{ij}(h_i) = \varepsilon_i \varphi_{ij}(h_i) + c_i \tag{5.1}$$

while an off-diagonal block A_{ik} has all its entries equal to some fixed rational integer c_{ik}.

Table 3.2

	1	$\leftarrow H_1^{\#} \rightarrow$		$\leftarrow H_s^{\#} \rightarrow$
1	1	$1 \cdots 1$	$\cdots$	$1 \cdots 1$
$\chi_{1j}, 1 \leqslant j \leqslant t_1$ $\{$		A_{11}		A_{1s}
			$\ddots$	
$\chi_{sj}, 1 \leqslant j \leqslant t_s$ $\{$		A_{s1}		A_{ss}

Suppose that some $c_{ik} = 0$. Let π_i be the character of N_i afforded by the regular representation of N_i/H_i. Although in general we cannot hope to unravel the induced generalised character $(\pi_i - \varphi_{i1})^G$, we can still get enough information to establish the following.

Lemma 22. *If* $c_{ik} = 0$, *then* $c_{ki} \neq 0$.

Proof. Assume that $c_{ik} = c_{ki} = 0$. Let $\theta = (\pi_i - \varphi_{i1})^G$. By the Frobenius reciprocity theorem restricting to N_i, $(\theta, 1_G) = 1$ and $(\theta, \chi_{kl}) = 0$ for each l so that, with integer coefficients,

$$\theta = 1_G + \sum_j a_{ij}\chi_{ij} + \sum_{m \neq i,k} \sum_n a_{mn}\chi_{mn}.$$

However, for each $m \neq i$, the exceptional characters occur in complex conjugate pairs $\chi_{mn}, \chi_{mn'}$ taking the same value on $H_i^{\#}$. Hence

$$a_{mn} = (\theta, \chi_{mn}) = ((\pi_i - \varphi_{i1}), \chi_{mn}|_{H_i}) = a_{mn'}.$$

If $g \in H_k^{\#}$,

$$\theta(g) = 1 + \sum_{m \neq i,k} a_{mn}c_{mk}.$$

In particular, $\theta(g)$ is an odd integer. On the other hand, since $\pi_i - \varphi_{i1}$ vanishes on $N_i - H_i$, we see that $\theta(g) = 0$, a contradiction. So $c_{ki} \neq 0$.

We conclude the proof of Theorem 20 by applying the orthogonality relations to obtain some bounds and then counting elements in order to obtain a contradiction.

We shall order the Hall subgroups so that $H = H_s$ has the smallest order and so that the blocks $A_{s1}, \ldots, A_{sq}$ are zero but $A_{s,q+1}, \ldots, A_{s,s-1}$ are not. Put $h = h_s$, $e = e_s$ and $t = t_s$. Also let $\Theta = \chi_{s1}$.

For $i \leqslant q$, we have $c_{is} \neq 0$. Hence, if $g \in H^{\#}$,

$$\sum_{j=1}^{t_i} |\chi_{ij}(g)|^2 \geqslant t_i.$$

The values $\chi_{sj}(g)$ are given by (5.1). Then, by applying column orthogonality in $N(H)$, we obtain the inequality

$$\begin{aligned}\sum_{j=1}^{t} |\chi_{sj}(g)|^2 &= \sum_{j=1}^{t} |\varepsilon\varphi_{sj}(g) + c_s|^2 \\ &\geqslant \sum_{j=1}^{t} |\varphi_{sj}(g)|^2 - |c_s| \sum_{j=1}^{t} (\varphi_{sj}(g) + \overline{\varphi_{sj}(g)}) + tc_s^2 \\ &\geqslant |C_H(g)| - e - 2|c_s| + tc_s^2 \\ &\geqslant |C_H(g)| - e,\end{aligned}$$

and a similar application in G yields the inequality

$$1 + \sum_{i=1}^{q} t_i \leqslant e. \tag{5.2}$$

Row orthogonality in $N(H)$ together with the values (5.1) give

$$\sum_{g \in H^{\#}} |\Theta(g)|^2 = eh - e^2 - 2c_s e + (h-1)c_s^2$$

whence, since $2e \leqslant h - 1$,

$$\sum_{g \in H^{\#}} |\Theta(g)|^2 \geqslant e(h-e). \tag{5.3}$$

Next we compute $\|\Theta\|^2$. Since each Hall subgroup H_i has $|G|/h_i e_i$ conjugates, we have the equation

$$\|\Theta\|^2 = |G|^{-1}\left(\Theta(1)^2 + \sum_{i=1}^{s} \frac{|G|}{h_i e_i} \sum_{g \in H_i^{\#}} |\Theta(g)|^2\right).$$

Put $\Theta(1) = d$. For $i = q+1, \ldots, s-1$, if $g \in H_i^{\#}$, then $|\Theta(g)|^2 \geqslant 1$. Hence we obtain the inequality

$$d^2 + \sum_{i=q+1}^{s-1} \frac{|G|}{h_i e_i}(h_i - 1) + \frac{|G|}{he} e(h-e) \leqslant |G|, \tag{5.4}$$

using (5.3). Counting elements in G, we see that

$$1 + \sum_{i=1}^{s} \frac{|G|}{h_i e_i}(h_i - 1) = |G|.$$

Substituting for the sum in (5.4), this reduces that inequality to

$$\sum_{i=1}^{q} \frac{t_i}{h_i} + \frac{h - 1 + e^2}{he} \geqslant 1 + \frac{d^2 - 1}{|G|} > 1,$$

the latter inequality since $d > 1$. Since $h < h_i$ for $i = 1, \ldots, s-1$, this inequality together with (5.2) implies that

$$\frac{e-1}{h}+\frac{h-1+e^2}{he}>1$$

or, putting $h-1=et$, that

$$(e-1)(t-2)<0.$$

Since $e>1$ and $t\geqslant 2$, this is impossible.

When, in Section 4 of the next chapter, we come to consider CN-groups of odd order, we shall require a substitute for Theorem 16. But the essential argument is the same, the main complication being that the exceptional characters for a nonabelian Hall subgroup do not all have the same degree.

Exercise

1. Show that a nonabelian CA-group of odd order is a Frobenius group.

6 Groups with self-normalising cyclic subgroups

In the treatment of CA-groups in the previous section, as will also be the case with CN-groups, the role of the nonexceptional characters was marked by the fact that a simple count of the number of conjugacy classes showed that every nonprincipal character was an exceptional character for some Hall subgroup. In the proof of the solubility of groups of odd order, the corresponding subgroups are no longer T.I.-sets, but a deep group theoretical analysis in a minimal simple group of odd order shows that there is an important cyclic subgroup which links two nonconjugate maximal subgroups. Such subgroups are 'self-normalising' in the following sense.

Let G be a finite group. A cyclic subgroup W is a *self-normalising cyclic subgroup* (with respect to a pair of nonidentity subgroups W_1 and W_2 of W) if $W=W_1\times W_2$ and $W=N_G(A)$ for each nonempty subset A of $V=W-(W_1\cup W_2)$. In particular, V is a T.I.-set in G and forms a closed set of special classes. This definition depends on W_1 and W_2: the subgroup W need not necessarily be self-normalising with respect to a different factorisation. What we can show is that, in the sense of Section 2, a natural induction occurs under character induction from $\mathscr{M}_W(V)$ to $\mathscr{C}\mathscr{h}(G)$ if W has odd order.

Theorem 23. *Let G be a finite group and let W be a self-normalising cyclic subgroup of odd order with respect to subgroups W_1 and W_2. Let $V=W-(W_1\cup W_2)$. Then the isometry from $\mathscr{M}(V)$ into $\mathscr{C}\mathscr{h}(G)$ afforded by character induction can be extended to a $\mathbb{Z}$-linear isometry*

$$\tau:\mathscr{C}\mathscr{h}(W)\rightarrow\mathscr{C}\mathscr{h}(G).$$

For $\alpha \in \mathscr{C}\mathscr{h}(W)$,

$$\alpha^\tau(v) = \alpha(v)$$

for all $v \in V$, *and any irreducible character of G not in the image of* τ *vanishes on* V.

The proof of this given by Feit and Thompson (1963; Lemma 13.1) depends on a careful analysis of character values and the fields in which they lie; here we shall prove Theorem 23 using the ideas of this chapter. In fact it is not even necessary to assume that W be cyclic or have odd order; a far weaker hypothesis will suffice. (See Collins (1988).)

As we have seen, in the theory of exceptional characters one usually cannot distinguish between characters and their negatives: so, to avoid carrying unknown signs, throughout this section we shall use the term 'character' for either an irreducible character or its negative, and the term 'distinct characters' for a set of such characters which are pairwise orthogonal. Furthermore, the use of distinct symbols will always imply that the characters are distinct in this sense.

We fix the following notation. Let $|W_1| = m+1$ and $|W_2| = n+1$. Let $\varphi_1, \ldots, \varphi_m$ be the nonprincipal characters of W which have W_2 in their kernels, and let $\psi_1, \ldots, \psi_n$ be those with W_1 in their kernels. For each pair i, j, let $\theta_{ij} = \varphi_i \psi_j$: then all the irreducible characters of W have been given. Put

$$\alpha_{ij} = 1_W - \varphi_i - \psi_j + \theta_{ij}, \quad 1 \leqslant i \leqslant m, 1 \leqslant j \leqslant n.$$

Lemma 24. *The generalised characters* α_{ij} *form a* $\mathbb{Z}$*-basis for* $\mathscr{M}(V)$.

Proof. $\alpha_{ij} = (1_W - \varphi_i)\cdot(1_W - \psi_j)$: hence $\alpha_{ij} \in \mathscr{M}(V)$. Now $\mathscr{M}(V)$ has $\mathbb{Z}$-rank mn. By suitably reordering characters, the relation matrix described in Section 2 for the α_{ij}s in terms of the irreducible characters of W has the form

$$A = (A_0 \;\; I):$$

in particular, the elementary divisors are all 1 so that the α_{ij}s generate a direct summand of $\mathscr{C}\mathscr{h}(W)$ of rank mn. Thus they form a basis for $\mathscr{M}(V)$.

Let σ denote the isometry from $\mathscr{M}(V)$ to $\mathscr{C}\mathscr{h}(G)$ afforded by character induction. We shall write $\tilde{\alpha}_{ij}$ for the image of α_{ij} under σ. The following is easily verified, using the Frobenius reciprocity theorem to establish part (i).

Lemma 25. (i) $\tilde{\alpha}_{ij}$ *involves* 1_G *with multiplicity* $+1$.

(ii) $(\tilde{\alpha}_{ij}, \tilde{\alpha}_{kl}) = 1 + \delta_{ik} + \delta_{jl} + \delta_{ik}\delta_{jl}$: *in particular,* $\|\tilde{\alpha}_{ij}\|^2 = 4$.

(iii) $\tilde{\alpha}_{ij}$ *is multiplicity-free.*

(iv) *If* $g \in G$ *and* g *is not conjugate to any element of* V, *then* $\tilde{\alpha}_{ij}(g) = 0$.

We now attempt to determine decompositions of the type (2.1) for the generalised characters $\tilde{\alpha}_{ij}$. To simplify notation, we shall use distinct natural numbers just to distinguish distinct suffixes i and similarly (and independently) for j, rather than to denote fixed values. Also, any statement for suffixes i will have its counterpart for the suffixes j, unless explicit assumptions have been made about m and n which remove symmetry.

Lemma 26. *$\tilde{\alpha}_{11}$ and $\tilde{\alpha}_{21}$ have exactly one nonprincipal constituent in common, with the same sign.*

Proof. If not, then Lemma 25(ii) forces the existence of characters χ_1, χ_2, and χ_3 such that

$$\tilde{\alpha}_{11} = 1_G + \chi_1 + \chi_2 + \chi_3$$

and

$$\tilde{\alpha}_{21} = 1_G + \chi_1 + \chi_2 - \chi_3.$$

But then $\chi_3(1) = 0$, which is impossible.

We remark, before proceeding further, that it is useful to think of the suffixes i and j as labelling the generalised characters α_{ij} presented in an $m \times n$ array. Then terms like 'hooks' and 'squares' in this array will have an obvious meaning.

Lemma 27. *$\tilde{\alpha}_{11}, \tilde{\alpha}_{21}$ and $\tilde{\alpha}_{12}$ do not share a common nonprincipal constituent.*

Proof. By symmetry, we may suppose that $m > n$. Then $m \geqslant 4$. Assume that the conclusion is false and that χ_1 is common constituent for $\tilde{\alpha}_{11}, \tilde{\alpha}_{21}$ and $\tilde{\alpha}_{12}$, necessarily with the same sign by Lemma 26. Then we may choose notation so that we have decompositions

$$\left.\begin{aligned} &\tilde{\alpha}_{11} = 1_G + \chi_1 + \chi_2 + \chi_3, \quad \tilde{\alpha}_{12} = 1_G + \chi_1 - \chi_2' + \chi_3'', \\ &\alpha_{21} = 1_G + \chi_1 + \chi_2' + \chi_3'. \end{aligned}\right\} \tag{6.1}$$

(This is a 'hook' in the array of $\tilde{\alpha}_{ij}$s.)

Case 1: $\tilde{\alpha}_{22}$ involves χ_1 ('square' with a common constituent).

We may choose χ_2 and χ_3 in (6.1) so that

$$\tilde{\alpha}_{22} = 1_G + \chi_1 - \chi_2 + \chi_3'''. \tag{6.2}$$

Notice that there is complete symmetry between $i = 1, 2$ and $j = 1, 2$.

Suppose, first, that some generalised character $\tilde{\alpha}_{31}$ involves χ_1. Then Lemma 25(ii) forces the decomposition

$$\tilde{\alpha}_{31} = 1_G + \chi_1 - \chi_3'' - \chi_3'''$$

and it is then impossible to find a decomposition for $\tilde{\alpha}_{41}$.

We may therefore assume, given (6.1) and (6.2), that *no* $\tilde{\alpha}_{ij}$, for $i \geqslant 3$ and $j \leqslant 2$, can involve χ_1. Suppose, next, that $\tilde{\alpha}_{31}$ involves χ_2 (which is distinguished from χ_3 by $(\tilde{\alpha}_{22}, \chi_2) \neq 0$). Then we are forced into decompositions

$$\tilde{\alpha}_{31} = 1_G + \chi_2 + \chi_3' + \chi_3''', \quad \tilde{\alpha}_{32} = 1_G + \chi_3'' + \chi_3''' + \chi_4.$$

Since $m \geqslant 4$ and $\tilde{\alpha}_{41} \neq \tilde{\alpha}_{31}$, a decomposition of the form

$$\tilde{\alpha}_{41} = 1_G + \chi_3 + \chi_3' + \chi_5$$

is forced, and it is impossible to find a decomposition for $\tilde{\alpha}_{42}$: this follows from an exhaustive analysis using Lemma 25(ii). Now, by symmetry, we may suppose that no $\tilde{\alpha}_{ij}$, for $i \geqslant 3$ and $j \leqslant 2$, involves either χ_2 or χ_2'. Then

$$\tilde{\alpha}_{31} = 1_G + \chi_3 + \chi_3' + \chi_4 \quad \text{and} \quad \tilde{\alpha}_{32} = 1_G + \chi_3'' + \chi_3''' + \chi_4,$$

and it is impossible to find a decomposition for $\tilde{\alpha}_{41}$.

Case 2: $\tilde{\alpha}_{22}$ does not involve χ_1.

Lemmas 25(ii) and 26 force a decomposition

$$\tilde{\alpha}_{22} = 1_G + \chi_3' + \chi_3'' + \chi_3'''.$$

In view of the completion of Case 1, we may assume that no 'squares' can share a common nonprincipal constituent. If $\tilde{\alpha}_{31}$ involves χ_1, then we can force

$$\tilde{\alpha}_{31} = 1_G + \chi_1 - \chi_3'' + \chi_3''',$$

whereupon no decomposition can be found for $\tilde{\alpha}_{32}$. So we may suppose that *no* $\tilde{\alpha}_{i1}$ involves χ_1 for $i \geqslant 3$, and we can force

$$\tilde{\alpha}_{31} = 1_G + \chi_2 + \chi_3' - \chi_3'''$$

(distinguishing χ_2 from χ_3 only by $(\tilde{\alpha}_{31}, \chi_2) \neq 0$). Now the argument above, applied to $\tilde{\alpha}_{21}, \tilde{\alpha}_{31}, \tilde{\alpha}_{41}$ and $\tilde{\alpha}_{22}$ as a 'long hook' yields an immediate contradiction.

Lemma 28. $\tilde{\alpha}_{11}$ *and* $\tilde{\alpha}_{22}$ *have no common nonprincipal constituent.*

Proof. Since we are assuming the absence of hooks with a common nonprincipal constituent, $\tilde{\alpha}_{22}$ cannot have in common with $\tilde{\alpha}_{11}$ either of its common nonprincipal constituents with $\tilde{\alpha}_{12}$ or $\tilde{\alpha}_{21}$. But then the existence of any common constituent will imply that $(\tilde{\alpha}_{11}, \tilde{\alpha}_{22}) = 0$ or 2, or that $\|\alpha_{22}\|^2 > 4$, contrary to Lemma 25(ii).

Lemma 29. *Assume that* $m \geqslant 4$. *Then there exist characters* $\{\Phi_i, \Theta_{i1}, \Theta_{i2}: i = 1, \ldots, m\}$, Ψ_1 *and* Ψ_2 *of* G *such that, for* $1 \leqslant i \leqslant m$,

$$\tilde{\alpha}_{i1} = 1_G + \Phi_i + \Psi_1 + \Theta_{i1} \quad \textit{and} \quad \tilde{\alpha}_{i2} = 1_G + \Phi_i + \Psi_2 + \Theta_{i2}.$$

Proof. By Lemma 28, we may start with decompositions

$$\tilde{\alpha}_{11} = 1_G + \Phi_1 + \Psi_1 + \Theta_{11}, \quad \tilde{\alpha}_{12} = 1_G + \Phi_1 + \Psi_2 + \Theta_{12},$$
$$\tilde{\alpha}_{21} = 1_G + \Phi_2 + \Psi_1 + \Theta_{21}, \quad \tilde{\alpha}_{22} = 1_G + \Phi_2 + \Psi_2 + \Theta_{22}.$$

If $(\tilde{\alpha}_{31}, \Psi_1) = 0$, then there exists a character Φ_3 such that

$$\tilde{\alpha}_{31} = 1_G + \Phi_3 + \Theta_{11} + \Theta_{21}.$$

Since there are no hooks, $(\tilde{\alpha}_{32}, \Phi_3) = 1$, but no decomposition can then be found for $\tilde{\alpha}_{41}$.

Similarly, we may suppose that $(\tilde{\alpha}_{32}, \Psi_2) \neq 0$, and we may label characters so that

$$\tilde{\alpha}_{31} = 1_G + \Phi_3 + \Psi_1 + \Theta_{31}, \quad \tilde{\alpha}_{32} = 1_G + \Phi_3 + \Psi_2 + \Theta_{32}.$$

We can now complete the proof of Theorem 23. We may suppose that $m > n$. Notice that both m and n must be even. By Lemma 29 we certainly have decompositions of the form

$$\tilde{\alpha}_{i1} = 1_G + \Phi_i + \Psi_1 + \Theta_{i1}, \quad \tilde{\alpha}_{i2} = 1_G + \Phi_i + \Psi_2 + \Theta_{i2}$$

for $i = 1, \ldots, m$. If a natural induction does not take place, then $n \geqslant 4$ and we may suppose that

$$\tilde{\alpha}_{13} = 1_G + \Theta_{11} + \Theta_{12} + \Psi_3$$

for some suitable ordering of the characters, following the argument of Lemma 29 (with m and n interchanged). But now Lemma 26 implies that $(\tilde{\alpha}_{i3}, \Psi_3) = 1$. By Lemma 29, $n \leqslant 3$, which is a contradiction. So a natural induction does take place and Theorem 13 implies that the conclusion of Theorem 23 holds, by writing

$$\tilde{\alpha}_{ij} = 1_G + \Phi_i + \Psi_j + \Theta_{ij}, \quad 1 \leqslant i \leqslant m, \ 1 \leqslant j \leqslant n,$$

and putting

$$\varphi_i^\tau = -\Phi_i, \quad \psi_j^\tau = -\Psi_j, \quad \theta_{ij}^\tau = \Theta_{ij}.$$

Exercises

1. Show that Theorem 23 can be extended to abelian groups W for which the factorisation $W = W_1 \times W_2$ has factors of coprime order.
2. Carry out the analogous analysis under the assumption that $|W|$ is divisible by 8.
3. Show that the conclusion of Theorem 23 holds if W has twice odd order.
4. Analyse the shapes of the 'unnatural' inductions that might occur if $|W| = 12$.

   [Hint. Part of the analysis is already contained in the proof of Lemma 27.]

4

Coherence and exceptional characters

Our goal in this chapter will be to develop methods which can be applied to study two classical problems which occurred early in the classification work on finite simple groups, those of CN-groups of odd order and Zassenhaus groups. In each case, the underlying theme is a simple group G possessing a subgroup N which is a Frobenius group with Frobenius kernel H a T.I.-set in G. The methods of the previous chapter are too elementary in themselves to handle these problems, primarily because they are dependent on having a basis for $\mathscr{M}_N(H^\#)$ which consists of generalised characters of sufficiently small norm that the induced characters can be directly decomposed.

We shall start by discussing some ideas due originally to Feit (1960) in his work on Zassenhaus groups and then more recent work of Sibley (1976). (See also [F; §31].) This will enable us to generalise the theorems of the last chapter. We shall then apply Sibley's results (as he himself observed was possible) to give refinements of the original proofs of the two classical problems to which we referred above. We emphasise that these are our target, so that the results of Feit and Sibley will not be presented in their strongest form. The methods of Feit were developed, in part, for application in the proof of the solubility of groups of odd order: for that, Suzuki's concept of special classes was insufficient, and tamely imbedded subsets were introduced. We shall not discuss these. Sibley's results can be strengthened using block theory and, again, we do not wish to become involved with that. However, all the new ideas that are due to Sibley are discussed in Section 3, and the interested reader can refer to Sibley's original papers (1975, 1976).

1 Coherence

In this section, we make no assumptions about the groups under consideration other than the stated hypotheses: as motivation, and this will be the situation in which we apply the results of this section, we may consider the situation where the given subgroup is a Frobenius group and its Frobenius kernel is a T.I.-set, thus generalising the situations considered in the previous chapter.

Let $\mathscr{S}$ be a given set of characters of a group (which will be clear from the context). Initially we do not assume that the members of $\mathscr{S}$ are necessarily irreducible, but only that they are positive linear combinations of irreducible characters. Let $\mathbb{Z}(\mathscr{S})$ denote the set of all integral linear combinations of characters in $\mathscr{S}$, and let

$$\mathbb{Z}_0(\mathscr{S}) = \{\alpha \in \mathbb{Z}(\mathscr{S}) \mid \alpha(1) = 0\}.$$

Now assume that N is a subgroup of a group G, and let $\mathscr{S}$ be a set of characters of N such that $\mathbb{Z}_0(\mathscr{S}) \neq \{0\}$. Suppose that there exists a linear isometry

$$\tau: \mathbb{Z}_0(\mathscr{S}) \to \mathscr{C}\mathscr{h}(G)$$

such that $\alpha^\tau(1) = 0$ for all $\alpha \in \mathbb{Z}_0(\mathscr{S})$. Then the pair $(\mathscr{S}, \tau)$ is *coherent* if τ can be extended to an isometry

$$\tau: \mathbb{Z}(\mathscr{S}) \to \mathscr{C}\mathscr{h}(G).$$

It need not be assumed that the extension should be uniquely determined. If the original map τ is understood from the context (for example, induction to G, although the *extension* will in general not be induction), we shall simply call $\mathscr{S}$ a *coherent set of characters.* The first two propositions relate the concept of coherence to the ideas studied in the previous chapter.

Proposition 1. *Assume that $\mathscr{S}$ consists of a set of irreducible characters of N and suppose that $(\mathscr{S}, \tau)$ is coherent. If $\chi \in \mathscr{S}$, then one of χ^τ and $-\chi^\tau$ is an irreducible character of G.*

Proof. Since $\chi^\tau \in \mathscr{C}\mathscr{h}(G)$ and $\|\chi^\tau\|_G = \|\chi\|_N = 1$, this is obvious.

In the situation of Proposition 1, the set of irreducible characters of G which are obtained in this way will be called the *exceptional characters* of G associated with the set of irreducible characters of N which lie in $\mathscr{S}$. We observe that, although the extension of τ from $\mathbb{Z}_0(\mathscr{S})$ to $\mathbb{Z}(\mathscr{S})$ need not be uniquely determined, the *set* of exceptional characters is characterised as the set of irreducible characters of G which occur as the constituents of the generalised characters α^τ for $\alpha \in \mathbb{Z}_0(\mathscr{S})$.

In the context of a subgroup N which is a Frobenius group whose kernel is a T.I.-set, we can see immediately that we cannot be using the whole of $\mathscr{M}_N(H^\#)$ as defined in Chapter 3 when discussing coherence since we cannot expect to achieve an isometry if $1_N \in \mathscr{S}$. Thus we will have no use here for the character π of N coming from the regular representation of N/H, as was discussed in Section 3.4. However, in the situation discussed there where H was abelian, we could take $\mathscr{S}$ to be the set of all characters of N for which H was not contained in the kernel. For H nonabelian, this is essentially our target although, as will be seen, it cannot quite be achieved.

As a first step, and now we return to the general situation, we have an exact analogue of Theorem 3.10.

Proposition 2. *Let N be a subgroup of a group G, and suppose that $\mathscr{S}$ is a set of irreducible characters of N, all having the same degree. Assume that $|\mathscr{S}| \geqslant 2$ and that $\tau: \mathbb{Z}_0(\mathscr{S}) \to \mathscr{C}h(G)$ is an isometry such that $\alpha^\tau(1) = 0$ whenever $\alpha \in \mathbb{Z}_0(\mathscr{S})$. Then $(\mathscr{S}, \tau)$ is coherent. Furthermore, the extension of τ to $\mathbb{Z}(\mathscr{S})$ is uniquely determined unless $|\mathscr{S}| = 2$, in which case exactly two extensions exist.*

Proof. Suppose that $\mathscr{S} = \{\varphi_1, \ldots, \varphi_n\}$. Then, as in the proof of Theorem 3.10, we obtain irreducible characters $\chi^{(1)}, \ldots, \chi^{(n)}$ of G and a sign $\varepsilon = \pm 1$ such that

$$(\varphi_i - \varphi_j)^\tau = \varepsilon(\chi^{(i)} - \chi^{(j)}), \quad 1 \leqslant i, j \leqslant n,$$

and we take $\varphi_i^\tau = \varepsilon\chi^{(i)}$. As in Theorem 3.10, ε is uniquely determined unless $n = 2$.

We next want a method to combine coherent sets of characters so that their union is coherent. The main result, due to Feit, is Theorem 5, but first we need two lemmas.

Lemma 3. *Let N be a subgroup of a group G, and let $\mathscr{S} = \{\varphi_1, \ldots, \varphi_n\}$ be a set of irreducible characters of N. Assume that $n \geqslant 2$ and that $(\mathscr{S}, \tau)$ is coherent, where τ is an isometry*

$$\tau: \mathbb{Z}_0(\mathscr{S}) \to \mathscr{C}h(G)$$

such that $\alpha^\tau(1) = 0$ if $\alpha \in \mathbb{Z}_0(\mathscr{S})$. For $i = 1, \ldots, n$, let $\varphi_i(1) = n_i$, and assume that not all the degrees n_i are equal. Then the extension of τ to $\mathbb{Z}(\mathscr{S})$ is uniquely determined and there is a sign $\varepsilon = \pm 1$ such that the exceptional characters $\chi^{(1)}, \ldots, \chi^{(n)}$ associated with $\mathscr{S}$ satisfy $\chi^{(i)} = \varepsilon\varphi_i^\tau$: in particular,

$$(n_i\varphi_j - n_j\varphi_i)^\tau = \varepsilon(n_i\chi^{(j)} - n_j\chi^{(i)}), \quad 1 \leqslant i, j \leqslant n.$$

Proof. Without loss, we may suppose that $n_1 \neq n_2$. Then, since

$$(n_1\varphi_2 - n_2\varphi_1)^\tau = n_1\varphi_2^\tau - n_2\varphi_1^\tau,$$

φ_1^τ is uniquely determined by the inner product

$$(\varphi_1^\tau, (n_1\varphi_2 - n_2\varphi_1)^\tau) = -n_2.$$

Now, for $i \geqslant 2$, φ_i^τ is determined by

$$n_1\varphi_i^\tau = n_i\varphi_1^\tau + (n_1\varphi_i - n_i\varphi_1)^\tau,$$

and $\varphi_1^\tau,\ldots,\varphi_n^\tau$ are either all irreducible or all the negatives of irreducible characters of G since $(n_i\varphi_j^\tau - n_j\varphi_i^\tau)(1)=0$.

Lemma 4. *Let N be a subgroup of the group G and let $\mathscr{S}$ be a set of irreducible characters of N for which there exists an isometry*

$$\tau\colon \mathbb{Z}_0(\mathscr{S})\to \mathscr{C}\!\mathscr{h}(G)$$

such that $\alpha^\tau(1)=0$ if $\alpha\in\mathbb{Z}_0(\mathscr{S})$. Let $\mathscr{S}_1=\{\varphi_{11},\ldots,\varphi_{1m}\}$ and $\mathscr{S}_2=\{\varphi_{21},\ldots,\varphi_{2n}\}$ be disjoint subsets of $\mathscr{S}$ with $m,n\geqslant 2$, and assume that $(\mathscr{S}_1,\tau)$ and $(\mathscr{S}_2,\tau)$ are coherent. Then the following hold.

(i) *The sets of exceptional characters associated with $\mathscr{S}_1$ and $\mathscr{S}_2$ are disjoint.*
(ii) *If $\mathscr{S}_1\cup\{\varphi_{21}\}$ is coherent, then $\mathscr{S}_1\cup\mathscr{S}_2$ is coherent, and the extension of τ to $\mathscr{S}_1\cup\mathscr{S}_2$ is uniquely determined.*

Proof. Let $n_{ij}=\varphi_{ij}(1)$. By Proposition 2 or Lemma 3, there are exceptional characters $\chi_{11},\ldots,\chi_{1m}$ and $\chi_{21},\ldots,\chi_{2n}$ associated with $\mathscr{S}_1$ and $\mathscr{S}_2$ respectively, and signs $\varepsilon_i=\pm1$, such that

$$(n_{it}\varphi_{is}-n_{is}\varphi_{it})^\tau=\varepsilon_i(n_{it}\chi_{is}-n_{is}\chi_{it}) \tag{1.1}$$

with uniqueness except in the case of two characters of equal degree. If $\chi_{11}=\chi_{21}$, say, then the isometry τ yields a value

$$\begin{aligned}I&=((n_{12}\chi_{11}-n_{11}\chi_{12}),(n_{22}\chi_{21}-n_{21}\chi_{22}))_G\\&=(\varepsilon_1(n_{12}\varphi_{11}-n_{11}\varphi_{12}),\varepsilon_2(n_{22}\varphi_{21}-n_{21}\varphi_{22}))_N\\&=0\end{aligned}$$

for the given inner product, while a direct computation gives

$$I=n_{12}n_{22}+n_{11}n_{21}(\chi_{12},\chi_{22})>0.$$

Thus the sets of exceptional characters are disjoint.

Let $\mathscr{S}'_1=\mathscr{S}_1\cup\{\varphi_{21}\}$ and assume that $\mathscr{S}'_1$ is coherent. Then $|\mathscr{S}'_1|\geqslant 3$ so that the extension of τ to $\mathscr{S}'_1$ is unique. Suppose, in addition to the notation above, that in *this* extension $\varphi_{21}^\tau=\varepsilon_1\chi'_{21}$.

If $n=2$ and $n_{21}=n_{22}$, then we may choose $\varepsilon_2=\varepsilon_1$ in (1.1) without loss. Then, as

$$((\varphi_{22}-\varphi_{21}),(n_{11}\varphi_{21}-n_{21}\varphi_{11}))=-n_{11}$$

and

$$(\varepsilon_2(\chi_{22}-\chi_{21}),\varepsilon_1(n_{11}\chi'_{21}-n_{21}\chi_{11}))=n_{11}((\chi_{22},\chi'_{21})-(\chi_{21},\chi'_{21})),$$

we deduce that $(\chi_{21},\chi'_{21})=1$. Thus $\chi'_{21}=\chi_{21}$, and putting $\varphi_{22}^\tau=\varepsilon_2\chi_{22}$ determines the unique extension of τ to $\mathscr{S}_1\cup\mathscr{S}_2$.

If $n \geqslant 3$ or $n_{22} \neq n_{21}$, then the signs $\varepsilon_1, \varepsilon_2$ are uniquely determined, and

$$\begin{aligned} I_t &= ((n_{11}\chi'_{21} - n_{21}\chi_{11}), (n_{2t}\chi_{21} - n_{21}\chi_{2t})) \\ &= (\varepsilon_1(n_{11}\varphi_{21} - n_{21}\varphi_{11}), \varepsilon_2(n_{2t}\varphi_{21} - n_{21}\varphi_{2t})) \\ &= \varepsilon_1\varepsilon_2 n_{11} n_{2t}, \end{aligned}$$

while direct computation yields the result

$$I_t = n_{11}n_{2t}(\chi'_{21}, \chi_{21}) - n_{11}n_{21}(\chi'_{21}, \chi_{2t})$$

for all $t \geqslant 2$. It follows that $\chi'_{21} = \chi_{21}$ and that $\varepsilon_1 = \varepsilon_2$. Now the coherence of $\mathscr{S}_1 \cup \mathscr{S}_2$ is clear.

We are now in a position to prove the major theorem of this section.

Theorem 5. *Let N be a subgroup of a group G and let $\mathscr{S}$ be a set of irreducible characters of N. Suppose that*

$$\tau: \mathbb{Z}_0(\mathscr{S}) \rightarrow \mathscr{C}\mathscr{h}(G)$$

is an isometry such that $\alpha^\tau(1) = 0$ if $\alpha \in \mathbb{Z}_0(\mathscr{S})$. Assume that the following conditions hold.

(i) *$\mathscr{S}$ is a disjoint union of subsets $\mathscr{S}_1, \ldots, \mathscr{S}_k$. For $i = 1, \ldots, k$, the set $\mathscr{S}_i = \{\varphi_{ij} \mid j = 1, \ldots, n_i\}$ is coherent if $n_i > 1$.*

(ii) *For each i, j, let $n_{ij} = \varphi_{ij}(1)$. Then n_{11} divides n_{m1} for each $m = 2, \ldots, k$.*

(iii) *For each $m = 2, \ldots, k$,*

$$\frac{1}{n_{11}} \sum_{i=1}^{m-1} \sum_{j=1}^{n_i} n_{ij}^2 > 2n_{m1}.$$

(iv) $n_1 \geqslant 2$.

Then $\mathscr{S}$ is coherent and the extension of τ to $\mathbb{Z}(\mathscr{S})$ is uniquely determined except when $k = 1$, $n_1 = 2$ and $n_{11} = n_{12}$.

Proof. If $k = 1$, the question of uniqueness, or the lack of it, is settled by Proposition 2 and Lemma 3. For $k \geqslant 2$, Lemma 4 reduces the proof to the case $k = 2$, $n_2 = 1$. So, to simplify the notation, take

$$\mathscr{S}_1 = \{\varphi_1, \ldots, \varphi_n\}, \quad \mathscr{S}_2 = \{\psi\}, \quad \varphi_i(1) = l_i \quad \text{and} \quad \psi(1) = d\varphi_1(1).$$

We first observe that condition (iii) prevents the possibility that $n = 2$ and $l_1 = l_2$. So the extension of τ to $\mathscr{S}_1$ is uniquely determined and there are characters $\chi^{(1)}, \ldots, \chi^{(n)}$ of G and a sign $\varepsilon = \pm 1$ such that

$$(l_i\varphi_1 - l_1\varphi_i)^\tau = \varepsilon(l_i\chi^{(1)} - l_1\chi^{(i)}), \quad i = 2, \ldots, n.$$

In particular,

$$l_i\chi^{(1)}(1) - l_1\chi^{(i)}(1) = 0,$$

so that

$$\frac{\chi^{(i)}(1)}{\chi^{(1)}(1)} = \frac{l_i}{l_1}:$$

thus $\sum_{i=1}^{n} a_i\varphi_i \in \mathbb{Z}_0(\mathscr{S})$ whenever $\sum_{i=1}^{n} a_i\chi^{(i)}(1) = 0$ for integers $a_1, \ldots, a_n$.

Now let $\theta = d\varphi_1 - \psi$. Then $\theta \in \mathbb{Z}_0(\mathscr{S})$. If there were to exist integers $a_1, \ldots, a_n$ such that

$$\theta^\tau = \sum_{i=1}^{n} a_i\chi^{(i)},$$

then, by the above,

$$\sum_{i=1}^{n} a_i\varphi_i \in \mathbb{Z}_0(\mathscr{S}),$$

and

$$\left\| \theta - \sum_{i=1}^{n} a_i\varphi_i \right\|_N = \left\| \theta^\tau - \varepsilon\left(\sum_{i=1}^{n} a_i\varphi_i \right)^\tau \right\|_G = \left\| \theta^\tau - \sum_{i=1}^{n} a_i\chi^{(i)} \right\|_G = 0.$$

So $\theta \in \mathbb{Z}_0(\mathscr{S}_1)$, which is false. Thus there is an irreducible character χ of G, distinct from $\chi^{(1)}, \ldots, \chi^{(n)}$, such that $(\theta^\tau, \chi) \neq 0$.

We would, of course, like to show that

$$\theta^\tau = \varepsilon(\mathrm{d}\chi^{(1)} - \chi),$$

for then we can put $\psi^\tau = \varepsilon\chi$ to give an extension of τ to $\mathbb{Z}_0(\mathscr{S})$. Certainly $\|\theta^\tau\|_G^2 = \|\theta\|_N^2 = d^2 + 1$, so let a be the nonnegative integer defined by $(\theta^\tau, \chi^{(1)}) = \varepsilon(d - a)$. It will suffice to show that $a = 0$.

If we write

$$l_1\chi^{(i)} = l_i\chi^{(1)} - (l_i\chi^{(1)} - l_1\chi^{(i)})$$

for $i = 2, \ldots, n$, we see that

$$\begin{aligned}(l_1\chi^{(i)}, \theta^\tau) &= (l_i\chi^{(1)}, \theta^\tau) - ((l_i\chi^{(1)} - l_1\chi^{(i)}), \theta^\tau)\\ &= \varepsilon l_i(d - a) - (\varepsilon(l_i\varphi_1 - l_1\varphi_i), \theta)\\ &= \varepsilon l_i(d - a) - \varepsilon l_i d\\ &= -\varepsilon l_i a.\end{aligned}$$

Hence

$$(\chi^{(i)}, \theta^\tau) = -\frac{\varepsilon l_i a}{l_1}, \quad i = 2, \ldots, n.$$

Now

$$\|\theta^\tau\|^2 \geqslant \sum_{i=1}^{n} (\theta^\tau, \chi^{(i)}) + (\theta^\tau, \chi).$$

Since $\|\theta^\tau\|^2 = d^2 + 1$, this inequality becomes

$$0 \geqslant -2ad + \frac{a^2}{l_1^2}\sum_{i=1}^{n} l_i^2$$

whence, by condition (iii), if $a \neq 0$, then

$$0 > -2ad + 2a^2d = 2d(a^2 - a),$$

which is impossible since a is an integer. Thus $a = 0$.

In the proof, condition (iii) plays a crucial role. At the time that Feit proved this result (with a further minor restriction on the degrees of the characters φ_{ij}), it was hoped that it would merely proved to be a device to make this proof possible. In the event, it turns out that there is a genuine obstruction, namely the Suzuki groups which were discovered shortly after Feit's result in a context which we shall describe in Section 5. In these groups one can find a set of characters arising out of a T.I.-set situation which is not coherent, where condition (iii) is violated. However, this point will be clearer after we have applied Theorem 5 in the next section.

Exercises

1. Show that, in the situation of Theorem 3.17, the set of nonlinear irreducible characters of N together with the character which is the sum of the two linear characters of N form a coherent set of characters. Is the same true of the set of *all* irreducible characters of N?
2. Establish formal coherence statements for Theorem 3.23 and the situation encountered in the proof of Theorem 3.20.
3. Prove Proposition 2 carefully.
4. Let A be an abelian Hall subgroup of a group G and let $N = N_G(A)$. Suppose that $C_G(A) = A \times M$ and that N/M is a Frobenius group with Frobenius kernel $(A \times M)/M$. Suppose that A is a T.I.-set in G. Show that the set of irreducible characters of N having M but not A in their kernels satisfies the hypothesis of Proposition 2 with respect to character induction. Can similar statements be established for the corresponding situation with A replaced by a nilpotent Hall subgroup H (if not for *all* characters, then possibly for subsets of characters)?

2 Frobenius groups as normalisers of Hall subgroups

Throughout this section, we shall work under the following hypothesis without specific reference.

Hypothesis I. G is a finite group, H is a Hall subgroup of G, and $N = N_G(H)$. The following conditions are satisfied.

(i) N is a Frobenius group with Frobenius kernel H.
(ii) $[N:H] = e$ and $|H| \neq e + 1$.

(iii) $\mathscr{S}$ is the set of irreducible characters of N which do not have H in their kernel, and there is an isometry

$$\tau: \mathbb{Z}_0(\mathscr{S}) \to \mathscr{C}\mathscr{h}(G)$$

such that $\alpha^\tau(1) = 0$ for all $\alpha \in \mathbb{Z}_0(\mathscr{S})$.

If H is abelian, which in particular is the case if e is even, then all the characters in $\mathscr{S}$ have degree e and $\mathscr{S}$ is coherent by Proposition 2. Our main aim in this and the next section will be to find conditions which will ensure that $(\mathscr{S}, \tau)$ is coherent. The strongest results can be obtained when H is a T.I.-set; then the elements of $H^\#$ form a set of special classes and, since the characters in $\mathscr{S}$ vanish outside H, it follows from Theorem 3.9 that the map from $\mathscr{M}_N(H^\#)$ to $\mathscr{C}\mathscr{h}(G)$ given by induction is an isometry. In this case we shall obtain detailed information about the values of the exceptional characters.

First we shall obtain a general result, due originally to Feit.

Theorem 6. *Assume that G is a group satisfying Hypothesis I. Then either (S, τ) is coherent, or H is a nonabelian p-group for some prime p and $[H:H'] \leqslant 4e^2 + 1$.*

If H is a T.I.-set and τ is given by character induction, Sibley (1976) was able to establish coherence in the case of H a p-group with p odd, and we shall discuss this in the next section. In fact, by using block theory, Sibley was able to extend his arguments to permit H to have a centraliser of coprime order: that is, a subgroup M such that $C_G(H) = Z(H) \times M$ and N/M is a Frobenius group, but we shall not investigate this extension here. Feit (1960) also had this as his hypothesis, though under a T.I.-set assumption with the isometry provided by character induction. In that respect, our hypothesis is more restrictive, but we can still see all the essential points.

We shall start by attempting to establish coherence simply under Hypothesis I. We shall fail since such a conclusion would, in generality, be false. The Suzuki groups provide a class of counterexamples, with the subgroup H of Theorem 6 being a Sylow 2-subgroup. These groups were discovered by Suzuki (1962b) in the context of the determination of a certain class of permutation groups known as *Zassenhaus groups*. An application of the methods discussed in this chapter to that classification is given in Section 5.

However, at the same time as proving Theorem 6 as stated, we shall be able to make a start on the proof of Sibley's theorem and we shall take the opportunity to establish suitable notation. Theorem 6 will be proved in Lemmas 7–11, and then we shall specialise to the case where the Hall subgroup H is a T.I.-set and where the isometry is given by induction; in Theorem 13, we shall obtain some information about character values. This will be sufficient to establish the Suzuki groups as the counterexamples mentioned above.

In view of the remark immediately following Hypothesis I, we may suppose that H is nonabelian. Let

$$H = H_0 \supset H_1 \supset \cdots \supset H_m \supset H_{m+1} = 1$$

be an N-chief series for H containing H': suppose that $H' = H_{r+1}$. Note in particular that $H_m \subseteq Z(H)$. For $i = 0, \ldots, m$, let $\mathscr{S}_i$ be the set of irreducible characters of N which have H_{i+1}, but not H_i, in their kernels, and put

$$\mathscr{X}_j = \bigcup_{i=0}^{j} \mathscr{S}_i, \quad j = 0, \ldots, m.$$

We recall from Theorem 2.43 that any character in $\mathscr{S}$ is induced from a character of H, and we note that the characters of H are themselves obtained as the products of irreducible characters of the Sylow subgroups of H. In particular, every character in $\mathscr{S}$ has degree divisible by e and every character in $\mathscr{X}_r$ has degree exactly e. Thus, if $|\mathscr{X}_r| \geqslant 2$, the set $\mathscr{X}_r$ is coherent by Proposition 2.

Lemma 7. *If $|\mathscr{X}_r| = 1$, then H is a nonabelian 2-group with $[H:H'] = e + 1$.*

Proof. If $|\mathscr{X}_r| = 1$, then in particular H/H' is an N-chief factor of H and so is an elementary abelian p-group, irreducible under the action of N/H with $[H:H'] = e + 1$. Since $H' \neq 1$, e is odd and H is a 2-group.

We may now work under the following, stronger assumptions.

Hypothesis II. G satisfies Hypothesis I. Furthermore,

(iv) e is odd,
(v) H is nonabelian, and
(vi) $\mathscr{X}_r$ is coherent.

We shall apply Theorem 5 in the following form.

Lemma 8. *Assume that Hypothesis II holds. Assume, further, that for some $j \geqslant r$ there exists a coherent subset $\mathscr{X}$ of $\mathscr{S}$ which contains $\mathscr{X}_j$ and such that, for all $\varphi \in \mathscr{S}_{j+1}$,*

$$e^{-1} \sum_{\chi \in \mathscr{X}} \chi(1)^2 > 2\varphi(1).$$

Then $\mathscr{X} \cup \mathscr{X}_{j+1}$ and $\mathscr{X}_{j+1}$ are coherent.

Proof. $\mathscr{X}$ contains a character of degree e lying in $\mathscr{X}_r$, and every character in $\mathscr{S}_{j+1}$ has degree divisible by e. Thus Theorem 5 immediately implies that $\mathscr{X} \cup \{\varphi\}$ is coherent for any φ in $\mathscr{S}_{j+1}$. By repeated application of this argument, we see that $\mathscr{X} \cup \mathscr{S}_{j+1}$ is coherent. Now $\mathscr{X} \cup \mathscr{S}_{j+1} = \mathscr{X} \cup \mathscr{X}_{j+1}$ since $\mathscr{X}_j \subseteq \mathscr{X}$, while any subset of a coherent set is necessarily coherent.

We exploit this lemma to reduce the proof of Theorem 6 to the case where H is a nonabelian p-group for some odd prime p. We shall need the following bound on character degrees.

Lemma 9. *If* $\psi \in \mathscr{S}_{j+1}$, *then* $\psi(1) \leqslant e[H:H_{j+1}]^{1/2}$.

Proof. We may regard ψ as a character of the Frobenius group N/H_{j+2}, for which $H_{j+1}/H_{j+2} \subseteq Z(H/H_{j+2})$ since H_{j+1}/H_{j+2} is a minimal normal subgroup of N/H_{j+2}. Now the bound follows from Theorem 2.43.

Lemma 10. *Assume Hypothesis II and suppose, further, that for some* $j \geqslant r$, *the set* $\mathscr{X}_j$ *is coherent but* $\mathscr{X}_{j+1}$ *is not. Then* $[H:H_{j+1}] \leqslant 4e^2 + 1$.

Proof. By Lemma 8, there is a character ψ in $\mathscr{S}_{j+1}$ such that

$$e^{-1} \sum_{\chi \in \mathscr{X}_j} \chi(1)^2 \leqslant 2\psi(1). \tag{2.1}$$

Furthermore, putting $h = [H:H_{j+1}]$, Lemma 9 implies that

$$\psi(1) \leqslant e\sqrt{h}. \tag{2.2}$$

Viewing $\mathscr{X}_j$ as the set of characters of N/H_{j+1} which do not have H/H_{j+1} in their kernels, we have

$$\sum_{\chi \in \mathscr{X}_j} \chi(1)^2 = eh - e:$$

combining this with the inequalities (2.1) and (2.2), we see that

$$h - 1 \leqslant e\sqrt{h},$$

or that

$$h^2 - 2(2e^2 + 1)h + 1 \leqslant 0.$$

Thus

$$h \leqslant (2e^2 + 1) + ((2e^2 + 1)^2 - 1)^{1/2} < 4e^2 + 2,$$

and so

$$h \leqslant 4e^2 + 1.$$

Lemma 11. *Assume the hypothesis of Lemma* 10. *Then* H *is a nonabelian* p*-group with* $[H:H'] \leqslant 4e^2 + 1$, *and, if* p *is odd,* $j = 0$.

Proof. Since e is odd, an N-chief p-factor of H has order of the form $\lambda e+1$, where λ is odd if $p=2$ and even if p is odd. Since

$$(2e+1)(e+1)^2>(2e+1)^2>4e^2+1, \tag{2.3}$$

we deduce first from Lemma 10 that H/H_{j+1} cannot have more than one N-chief factor of odd order and then, since $H'\supseteq H_{j+1}$, that the entire conclusion holds if H has prime power order.

If H does not have prime power order, then each prime divisor of $|H|$ also divides $[H:H']$ so that H is a $\{2,p\}$-group for some odd prime p. Then the inequalities (2.3) force $j=1$ and $H'=H_2$ with H/H' having exactly two chief N-factors, of orders

$$e+1=2^s \quad \text{and} \quad 2e+1=p^t$$

for some $s,t\geqslant 1$. Now, by Lemma 9, if $\varphi\in\mathscr{S}_2$, then $\varphi(1)\leqslant e\cdot 2^{s/2}$ or $\varphi(1)\leqslant e\cdot p^{t/2}$, according to whether H_2/H_3 is a 2-group or a p-group for p odd. Then, since

$$e^{-1}\sum_{\chi\in\mathscr{X}_1}\chi(1)^2=e^{-1}(e(e+1)(2e+2)-e)=2e^2+3e$$

while

$$2\cdot 2^{s/2}<2p^{t/2}=2(2e+1)^{1/2}<2(e+1),$$

we see that

$$e^{-1}\sum_{\chi\in\mathscr{X}_1}\chi(1)^2>2\varphi(1)$$

for all $\varphi\in\mathscr{S}_2$, so that $\mathscr{X}_2$ is coherent by Lemma 8, contrary to our assertion above that $j=1$.

Since, if $\mathscr{S}$ is not coherent, the hypothesis of Lemma 10 must hold for some j, Theorem 6 follows immediately. For convenience, we make one further reduction in the general case.

Lemma 12. *In order to establish coherence under Hypothesis II, it is sufficient to show that* $\mathscr{S}_0\cup\mathscr{S}_m$ *is coherent.*

Proof. Suppose that $\mathscr{X}_j\cup\mathscr{S}_m$ is coherent for some $j\geqslant 0$. Put $\mathscr{X}=\mathscr{X}_j\cup\mathscr{S}_m$. We shall apply Lemma 8 to show that $\mathscr{X}_{j+1}\cup\mathscr{S}_m$ is coherent.

For $\varphi\in\mathscr{S}_{j+1}$, we know that $\varphi(1)\leqslant e[H:H_{j+1}]^{1/2}$ by Lemma 9. Now

$$\begin{aligned}\sum_{\chi\in\mathscr{X}_j\cup\mathscr{S}_m}\chi(1)^2&=\sum_{\chi\in\mathscr{X}_j}\chi(1)^2+e(|H|-[H:H_m])\\&=\sum_{\chi\in\mathscr{X}_j}\chi(1)^2+e[H:H_m](|H_m|-1)\\&\geqslant\sum_{\chi\in\mathscr{X}_j}\chi(1)^2+2e^2[H:H_m]\\&>2e^2[H:H_{j+1}]^{1/2}\geqslant 2e\varphi(1).\end{aligned}$$

Thus $\mathscr{X}_{j+1} \cup \mathscr{S}_m$ is coherent and hence, inductively, so is $\mathscr{S}$.

We note that the coherence of $\mathscr{S}_0 \cup \mathscr{S}_m$ ensures that the contributions of the squares of the degrees of the characters in $\mathscr{S}_m$ will provide the linkage between $\mathscr{X}_j$ and $\mathscr{S}_{j+1}$ at each stage, thus overcoming the obstacle that we used to carry out the reductions up to Lemma 11.

At this point, we must cease working in generality. For the remainder of this section and the next, we shall assume the following hypothesis.

Hypothesis III. G satisfies Hypothesis I, with H a T.I.-set in G and τ the isometry given by character induction.

In this case, Sibley's theorem establishes coherence when H is of odd order without restriction, but before we do so we shall obtain an analogue of Theorem 3.16 for character values when $(\mathscr{S}, \tau)$ is coherent.

Theorem 13. *Assume Hypothesis III and suppose that $(\mathscr{S}, \tau)$ is coherent. Then the following hold.*

(i) *There exists a rational integer c such that, for all $\varphi \in \mathscr{S}$,*

$$\varphi^\tau(h) = \varphi(h) + \frac{\varphi(1)}{e} \cdot c$$

for all $h \in H^\#$.

(ii) $\sum_{\varphi \in \mathscr{S}} |\varphi^\tau(h)|^2 \geqslant |C_G(h)| - e$ *for all $h \in H^\#$.*

(iii) *If $\varphi, \psi \in \mathscr{S}$, then $\varphi(1)\psi^\tau(1) = \psi(1)\varphi^\tau(1)$.*

(iv) *If χ is an irreducible character of G such that $\pm\chi \notin \mathscr{S}^\tau$, then χ is constant and rational-valued on $H^\#$.*

(v) *If $\varphi \in \mathscr{S}$, then φ^τ takes rational values on elements of G which are not conjugate to elements of H.*

Proof. Let $\mathscr{S} = \{\varphi_1, \ldots, \varphi_t\}$, and suppose that $H' \subseteq \ker \varphi_1$ so that $\varphi_1(1) = e$. Suppose that

$$\varphi_i^\tau|_N = \sum_{j=1}^{t} a_{ij}\varphi_j + \lambda_i, \quad i = 1, \ldots, t, \tag{2.4}$$

where $a_{ij} \in \mathbb{Z}$ and λ_i is an integral linear combination of characters of N which have H in their kernels. Then application of the Frobenius reciprocity theorem and the coherence of $(\mathscr{S}, \tau)$ yields a set of equations

$$\begin{aligned} \frac{\varphi_j(1)}{e} \cdot a_{i1} - a_{ij} &= \left(\varphi_i^\tau|_N, \left(\frac{\varphi_j(1)}{e}\varphi_1 - \varphi_j \right) \right)_N \\ &= \left(\varphi_i^\tau, \left(\frac{\varphi_j(1)}{e}\varphi_1 - \varphi_j \right)^\tau \right)_G \end{aligned}$$

$$=\left(\varphi_i^\tau,\left(\frac{\varphi_j(1)}{e}\varphi_1^\tau-\varphi_j^\tau\right)\right)_G$$

$$=\frac{\varphi_j(1)}{e}\delta_{i1}-\delta_{ij}.$$

Thus

$$a_{ij}=\frac{\varphi_j(1)}{e}(a_{i1}-\delta_{i1})+\delta_{ij},$$

so that (2.4) can be rewritten as

$$\varphi_i^\tau|_N=\varphi_i+\frac{a_{i1}-\delta_{i1}}{e}\sum_{j=1}^t\varphi_j(1)\varphi_j+\lambda_i.$$

Now an application of the orthogonality relations to N implies that

$$\sum_{j=1}^t\varphi_j(1)\varphi_j(h)=-e \tag{2.5}$$

if $h\in H^\#$, while $\lambda_i(h)=\lambda_i(1)$; hence there exist integers $c_1,\ldots,c_t$ such that, for $h\in H^\#$,

$$\varphi_i^\tau(h)=\varphi_i(h)+c_i,\quad i=1,\ldots,t.$$

However, by Theorem 3.9(i), if $h\in H^\#$ then

$$\left(\frac{\varphi_i(1)}{e}\varphi_1^\tau-\varphi_i^\tau\right)(h)=\left(\frac{\varphi_i(1)}{e}\varphi_1-\varphi_i\right)^G(h)$$

$$=\frac{\varphi_i(1)}{e}\varphi_1(h)-\varphi_i(h),$$

and so

$$c_i=\varphi_i^\tau(h)-\varphi_i(h)=\frac{\varphi_i(1)}{e}(\varphi_1^\tau(h)-\varphi_1(h))=\frac{\varphi_i(1)}{e}\cdot c$$

as required for part (i).

Now we may immediately compute that, if $h\in H^\#$,

$$\sum_{\varphi\in\mathscr{S}}|\varphi^\tau(h)|^2=\sum_{\varphi\in\mathscr{S}}\left(|\varphi(h)|^2+\frac{c}{e}(\varphi(h)+\overline{\varphi(h)})\varphi(1)+\frac{c^2}{e^2}\varphi(1)^2\right)$$

$$=|C_G(h)|-e-2c+\frac{|H|-1}{e}\cdot c^2.$$

Since $(|H|-1)/e\geqslant 2$, it follows that

$$\sum_{\varphi\in\mathscr{S}}|\varphi^\tau(h)|^2\geqslant|C_G(h)|-e.$$

Part (iii) is clear since $(\varphi(1)\psi-\psi(1)\varphi)\in\mathbb{Z}_0(\mathscr{S})$ for $\varphi,\psi\in\mathscr{S}$.

If χ is an irreducible character of G distinct from the exceptional characters associated with $\mathscr{S}$, then, by the Frobenius reciprocity theorem, for $1\leqslant i,j\leqslant t$,

$$(\chi|_N,(\varphi_j(1)\varphi_i-\varphi_i(1)\varphi_j))_N=(\chi,(\varphi_j(1)\varphi_i^\tau-\varphi_i(1)\varphi_j^\tau))$$

$$=0:$$

putting $j = 1$, it follows that

$$(\chi|_N, \varphi_i)_N = \frac{\varphi_i(1)}{\varphi_1(1)}(\chi|_N, \varphi_1)_N.$$

Hence, for some integer k and an integral linear combination λ of the irreducible characters of N which have H in their kernels,

$$\chi|_N = \frac{k}{\varphi_1(1)} \sum_{i=1}^{t} \varphi_i(1)\varphi_i + \lambda,$$

so that, by (2.5), for all $h \in H^{\#}$,

$$\chi(h) = -k + \lambda(1).$$

So (iv) holds.

Let $\varphi \in \mathscr{S}$. If $\varphi^\tau(g)$ were irrational for some element g not conjugate to an element of $H^{\#}$, then, as g would have order coprime to $|H|$, there would be an automorphism θ of some splitting field for G such that $(\varphi^\tau(g))^\theta \neq \varphi^\tau(g)$ but such that $(\varphi^\tau|_H)^\theta = \varphi^\tau|_H$. However, by (i), φ is not constant on $H^{\#}$ and nor then is φ^τ (see Exercise 4 of Section 2.6), so that (iv) implies that $(\varphi^\tau)^\theta \in \pm \mathscr{S}^\tau$; but then $(\varphi^\tau)^\theta = \varphi^\tau$ since φ^τ is characterised in $\mathscr{S}^\tau$ by $\varphi^\tau|_H$. This contradiction establishes (v).

We observe that a careful inspection of the proof will show that there is equality in (ii) if and only if either $c = 0$ or if $c = 1$ and $|H| = 2e + 1$. In the latter case, $|\mathscr{S}| = 2$ and the indeterminacy of Proposition 2 means that the other extension of the isometry τ to $\mathscr{S}$ will yield $c = 0$.

In Section 3 we will show that, under suitable restrictions on e, the constant c is zero. This was achieved in Theorem 3.17 by using the characters of N/H. Such methods, giving a complete natural induction, are not available in general, and we will need to resort to delicate class algebra constant calculations. However, Theorem 13 is sufficient to establish the following example of noncoherence, demonstrating the need for some restriction in Theorem 6 and its refinements.

Example. The Suzuki groups $\mathrm{Sz}(q)$ are defined as simple subgroups of the symplectic groups $\mathrm{Sp}_4(q)$ over fields of order q, where $q = 2^{2n+1}$ with $n \geqslant 1$. For our present purposes, we will need to assume only the following properties of such groups, although we should remark that these conditions formed part of the information obtained group theoretically on the way to their construction.

(I) $|\mathrm{Sz}(q)| = q^2(q-1)(q^2+1)$.

(II) Let $q = 2r^2$. The prime divisors of $|\mathrm{Sz}(q)|$ fall into four disjoint sets, $\{2\}$ and the prime divisors of $q-1$, $q+2r+1$ and $q-2r+1$, respectively. For each set, $\mathrm{Sz}(q)$ contains a nilpotent Hall subgroup H such that $N(H)$ is a Frobenius group with H as Frobenius kernel: furthermore, H is a T.I.-set.

(III) If $|H| = q \pm 2r + 1$, then H is cyclic and $[N(H):H] = 4$.

(IV) If $|H| = q - 1$, then H is cyclic and $[N(H):H] = 2$.

(V) A Sylow 2-subgroup S is nonabelian of order q^2, $|Z(S)| = q$ and every involution of S lies in $Z(S)$. All elements of $S - Z(S)$ have order 4, and centralisers of order $2q$. S is a T.I.-set in $\mathrm{Sz}(q)$ and $N(S)/S$ is cyclic of order $q - 1$.

Let H_1, H_2, H_3 be nilpotent Hall subgroups of orders $q - 1$, $q + 2r + 1$ and $q - 2r + 1$ respectively. Then every element of $\mathrm{Sz}(q)$ is conjugate to an element of S, H_1, H_2 or H_3, and it is easily checked that the nonidentity conjugacy classes of $\mathrm{Sz}(q)$ can be described as follows.

(VI) S contains one class of involutions and two conjugacy classes of elements of order 4, with such elements not conjugate to their inverses.
H_1 contains $\frac{1}{2}(q - 2)$ classes.
H_2 contains $\frac{1}{4}(q + 2r)$ classes.
H_3 contains $\frac{1}{4}(q - 2r)$ classes.

Let $N_i = N(H_i)$, $i = 1, 2, 3$. Then the methods of Section 3.4 will show that a natural induction occurs when considering the isometries

$$\tau_i: \mathscr{M}_{N_i}(H_i^{\#}) \to \mathscr{C}\mathscr{h}(\mathrm{Sz}(q))$$

given by induction, if $q \geqslant 32$. Special arguments are needed when $q = 8$, but the same conclusion can be drawn, except that there are no exceptional characters corresponding to the subgroup H_3 which has order 5 in this case.

Suppose, then, that $q \geqslant 32$. Let the nonlinear characters of N_i be denoted by $\{\varphi_j^{(i)}\}$, $i = 1, 2, 3$, and let the corresponding exceptional characters of $\mathrm{Sz}(q)$ be $\{\Phi_j^{(i)}\}$. Since the exceptional characters for each i all take the same values on nonspecial classes by Theorem 3.16, the exceptional characters $\Phi_j^{(i)}$ are all distinct and, further, there are too many exceptional characters for the exceptional characters corresponding to one cyclic Hall subgroup to be nonexceptional characters for another.

There are exactly four further irreducible characters for $\mathrm{Sz}(q)$, including the trivial character; let them be $\chi_0 = 1, \chi_1, \chi_2$ and χ_3, and suppose that χ_1 is the character obtained by the natural induction from N_1. Notice that all of χ_0, χ_1, χ_2 and χ_3 occur in the natural inductions from N_2 and N_3. Hence an analogue of Theorem 3.17 yields the following relations

Table 4.1

	1	2	4_1	4_2	odd
φ_1	1	1	1	1	1
φ_2	.	.	.	.	*
.	.	.	.	.	.
.	.	.	.	.	.
φ_{q-1}	1	1	1	1	*
ψ_1	$q-1$	$q-1$	-1	-1	0
ψ_2	$r(q-1)$	$-r$	ir	$-ir$	0
ψ_3	$r(q-1)$	$-r$	$-ir$	ir	0

and congruences on degrees:

(VII) (i) $\chi_1(1) \equiv \varepsilon \pmod{(q-1)}$ for some $\varepsilon = \pm 1$.
(ii) $1 + \varepsilon\chi_1(1) - \varepsilon\Phi_j^{(1)}(1) = 0$.
(iii) $\Phi_j^{(1)}(1) \equiv 0 \pmod{(q+2r+1)}$.
(iv) $\Phi_j^{(1)}(1) \equiv 0 \pmod{(q-2r+1)}$.

From (iii) and (iv), it follows that q^2+1 divides $\Phi_j^{(1)}(1)$, while from (i) and (ii), $(q-1, \Phi_j^{(1)}(1)) = 1$. Now, in addition,

$$\chi_1(1) \equiv \pm 1 \pmod{(q+2r+1)}.$$

If $\Phi_j^{(1)}(1)$ were even then, by (ii), $\chi_1(1)$ would be odd and hence $\chi_1(1) = 1$, contrary to the simplicity of $\mathrm{Sz}(q)$. Thus

$$\Phi_j^{(1)}(1) = q^2 + 1 \quad \text{and} \quad \chi_1(1) = q^2.$$

Now $\chi_1(h) = \pm 1$ for $h \in H_i^\#$, $i = 1, 2, 3$. Hence, using the orthogonality relation $\sum_g |\chi_1(g)|^2 = |\mathrm{Sz}(q)|$, we see that χ_1 must vanish on $S^\#$. (In fact, since $\chi_1(1)$ is divisible by the highest power of 2 which divides $|\mathrm{Sz}(q)|$, it is a general theorem that χ_1 lies in a so-called block of defect zero and will vanish on every element of even order. See Proposition 6.19.)

Now let $N = N(S)$ and consider the isometry from $\mathcal{M}_N(S^\#)$ to $\mathcal{C}\mathcal{h}(\mathrm{Sz}(q))$ given by induction. It is an easy exercise to show that, since the conjugacy classes of elements of order 4 in N are nonreal, N has the character table shown in Table 4.1.

Suppose that $\mathcal{S} = \{\psi_1, \psi_2, \psi_3\}$ were coherent, and let Ψ_1, Ψ_2, Ψ_3 be the associated exceptional characters. Then the argument above that showed that the characters $\Phi_j^{(i)}$ were all distinct shows them to be distinct from Ψ_1, Ψ_2, Ψ_3 also. From Theorem 13, we see that Ψ_1, Ψ_2 and Ψ_3 cannot be constant over the three classes of elements of order 2 or 4 so that none can be χ_1. Since that leaves only χ_2 and χ_3, this is impossible and thus $\mathcal{S}$ cannot be coherent. We observe that, in the general notation of this section corresponding to S we will have $|\mathscr{X}_r| = 1$. A similar analysis applies if $q = 8$.

Exercises

1. Continue the analysis above and show that the following is the character table of $\mathrm{Sz}(q)$, where, for $i = 1, 2, 3$, $h_i \in H_i^\#$, τ is an involution, and σ_1, σ_2 are nonconjugate elements of order 4.

	1	τ	σ_1	σ_2	h_1	h_2	h_3
χ_0	1	1	1	1	1	1	1
χ_1	q^2	0	0	0	1	-1	-1
χ_2	$r(q-1)$	$-r$	ir	$-ir$	0	1	-1
χ_3	$r(q-1)$	$-r$	$-ir$	ir	0	1	-1
$\Phi_j^{(1)}$	q^2+1	1	1	1	$\varphi_j^{(1)}(h_1)$	0	0
$\Phi_j^{(2)}$	$(q-1)(q-2r+1)$	$2r-1$	-1	-1	0	$-\varphi_j^{(2)}(h_2)$	0
$\Phi_j^{(3)}$	$(q-1)(q+2r+1)$	$-2r-1$	-1	-1	0	0	$-\varphi_j^{(3)}(h_3)$

[Note. χ_2 and χ_3 *are* the exceptional characters corresponding to the pair of charcters $\{\psi_2, \psi_3\}$ of $N(S)$.]

2. Decompose the induced character $(\psi_2 - r\psi_1)^G$.

3 Sibley's theorems

Throughout this section, we shall assume Hypothesis III of Section 2, namely, with the notation of that section, we shall be working under the assumption that H is a T.I.-set and that the isometry τ is given by character induction. We shall first prove the result due to Sibley (1976) that, under this hypothesis, the conclusion of Theorem 6 can be improved to give coherence except when $p = 2$.

Theorem 14. *Assume that G is a group satisfying Hypothesis III. Then either $(\mathscr{S}, \tau)$ is coherent, or H is a nonabelian 2-group and $[H:H'] \leqslant 4e^2 + 1$.*

In the case where $\mathscr{S}$ is not coherent, the Sylow 2-subgroups of G are T.I.-sets. Suzuki (1964) gave a complete description of such groups, and the only groups satisfying Hypothesis III with H a nonabelian 2-group and N a proper subgroup of G are in fact the Suzuki groups. Thus these groups provide the *only* counterexamples to coherence under this hypothesis.

In view of Theorem 6, in order to prove Theorem 14 we may assume that the following holds.

Lemma 15. *H is a nonabelian p-group for some odd prime p. In particular, e is odd and $|\mathscr{X}_r| \geqslant 2$.*

The proof now falls into three parts. First we must show that $\mathscr{S}_m$ is coherent. For the purposes of subsequent exposition, it is convenient to show that each $\mathscr{S}_i$ is coherent. We shall then obtain information about the values of the exceptional characters corresponding to $\mathscr{S}_0$ on $H_m^{\#}$. These steps use arguments already seen. Next, and this is the major innovation that Sibley introduced, we look at class algebra constants in G corresponding to conjugacy class sums for elements in H_m. This provides the information which, combined with the character values already obtained, forces the coherence of $\mathscr{S}_0 \cup \mathscr{S}_m$. Then Lemma 12 implies the coherence of $\mathscr{S}$.

Lemma 16. *No character in $\mathscr{S}$ is real. In particular, characters in each $\mathscr{S}_i$ appear in complex conjugate pairs.*

Proof. This is immediate since, by Lemma 15, we can assume that N has odd order so that no nonidentity element is real.

Lemma 17. *Each set $\mathscr{S}_i$ is coherent. If $i_1 \neq i_2$, the sets of exceptional characters corresponding to $\mathscr{S}_{i_1}$ and $\mathscr{S}_{i_2}$ are disjoint.*

Proof. Fix i; we shall break $\mathscr{S}_i$ into subsets of characters having the same degree and build $\mathscr{S}_i$ using Theorem 5. Let

$$ep^{k_1} < ep^{k_2} < \cdots < ep^{k_s}$$

be the distinct degrees of the characters in $\mathscr{S}_i$. Let

$$\mathscr{S}^{(j)} = \{\varphi \in \mathscr{S}_i \,|\, \varphi(1) = ep^{k_j}\}$$

and let $|\mathscr{S}^{(j)}| = n_j$. Then $n_j \geqslant 2$ for each j, and $\mathscr{S}^{(j)}$ is coherent by Proposition 2. Now suppose that $s \geqslant 2$. Then

$$\sum_{j=1}^{s} n_j e^2 p^{2k_j} = e[H{:}H_{i+1}] - e[H{:}H_i].$$

By Lemma 9, $[H{:}H_i]$ is divisible by p^{2k_s}. Hence, by successive descent, since $(e, p) = 1$ we obtain congruences

$$\sum_{j=1}^{r-1} n_j e^2 p^{2k_j} \equiv 0 \pmod{e^2 p^{2k_r}}$$

for each $r = 2, \ldots, s$. These congruences imply inequalities

$$\sum_{j=1}^{r-1} n_j e^2 p^{2(k_j - k_1)} \geqslant e^2 p^{2(k_r - k_1)} > 2e^2 p^{k_r - k_1}$$

since p is odd, so that $\mathscr{S}_i$ is coherent by Theorem 5. Now the disjointness of the sets of associated exceptional characters follows from Lemma 4(i).

The extension of τ to $\mathscr{S}_i$ is unique unless $|\mathscr{S}_i|=2$, in which case two extensions exist. We may then need to make a choice later, but we shall abuse notation and allow τ to denote any such extension, for each $\mathscr{S}_i$ *individually*.

For $i=0,\dots,m$, let ef_i be the least degree of the characters in $\mathscr{S}_i$, and let

$$\alpha_i = \sum_{\varphi\in\mathscr{S}_i} \frac{\varphi(1)}{ef_i}\varphi:$$

then $ef_i\alpha_i$ is the contribution from $\mathscr{S}_i$ to the regular character of N. In particular, $\alpha_i|_{H_{i+1}}$ is constant.

Lemma 18. *Let $\varphi\in\mathscr{S}_0$. Then there exist integers $c_0,\dots,c_m,d$, independent of φ, such that, for all $g\in H$,*

$$\varphi^\tau(g) = \varphi(g) + \sum_{i=0}^{m} c_i\alpha_i(g) + d.$$

Proof. We may certainly write, for each $\varphi\in\mathscr{S}_0$,

$$\varphi^\tau|_N = \varphi + \sum_{i=0}^{m}\sum_{\psi\in\mathscr{S}_i} a_{\varphi\psi}\psi + \lambda_\varphi \tag{3.1}$$

where λ_φ is a linear combination of characters of N whose kernels contain H. Now, if $\varphi,\varphi'\in\mathscr{S}_0$, then $\varphi(1)=\varphi'(1)=e$ and, by Theorem 3.9, $(\varphi-\varphi')^\tau|_H=(\varphi-\varphi')|_H$. Thus

$$\left(\sum_{i=0}^{m}\sum_{\psi\in\mathscr{S}_i}(a_{\varphi\psi}-a_{\varphi'\psi})\psi + (\lambda_\varphi-\lambda_{\varphi'})\right)\Bigg|_H \equiv 0.$$

It follows first that $a_{\varphi\psi}=a_{\varphi'\psi}$ for each ψ since the restrictions $\{\psi|_H\}$ and 1_H are linearly independent, and then that $\lambda_\varphi|_H=\lambda_{\varphi'}|_H$. Hence, in (3.1), the integers $a_{\varphi\psi}$ are independent of φ and we may take $d=\lambda_\varphi(1)$. Also, write $a_\psi=a_{\varphi\psi}$.

Let $i\geqslant 1$. If $\psi,\psi'\in\mathscr{S}_i$, then $\varphi^\tau\neq\psi^\tau,\psi'^\tau$ by Lemma 17. Hence, by the Frobenius reciprocity theorem,

$$\begin{aligned}(\varphi^\tau|_N,(\psi'(1)\psi-\psi(1)\psi'))_N &= (\varphi^\tau,(\psi'(1)\psi-\psi(1)\psi')^G)_G\\ &= (\varphi^\tau,(\psi'(1)\psi^\tau-\psi(1)\psi'^\tau))_G\\ &= 0.\end{aligned}$$

Thus

$$a_\psi\psi'(1) = a_{\psi'}\psi(1)$$

and hence, for $i\geqslant 1$, we may take c_i to be the coefficient of a character of

$\mathscr{S}_i$ of degree ef_i in (3.1). Similarly, since all characters in $\mathscr{S}_0$ have degree e and the corresponding exceptional characters are distinct, we see first that $a_{\varphi'}$ is constant for $\varphi' \neq \varphi$, and then that $a_\varphi = a_{\varphi'}$, and we take this value for c_0.

Corollary 19. *Let $\varphi \in \mathscr{S}_0$. Then, if $g \in H_m^\#$,*

$$\varphi^\tau(1) = \varphi^\tau(g) + \frac{c_m|H|}{f_m}.$$

Proof. By Lemma 18, since $H_m \subseteq \ker\varphi$ and $\alpha_i|_{H_m}$ is constant for $i < m$, we have

$$\varphi^\tau(1) - \varphi^\tau(g) = c_m(\alpha_m(1) - \alpha_m(g)).$$

Now

$$ef_m\alpha_m(g) + [N:H_m] = 0$$

and

$$ef_m\alpha_m(1) + [N:H_m] = |N|.$$

Hence

$$\alpha_m(1) - \alpha_m(g) = \frac{|H|}{f_m}.$$

We now consider class algebra constants. For $g, h \in H_m^\#$ and $x \in G$, let a_{ghx}^G denote the class algebra constants in $\mathbb{C}G$ defined by

$$C_g C_h = \sum_x a_{ghx}^G C_x$$

where C_g, C_h, C_x denote the conjugacy class sums for the classes containing g, h, x respectively. For $\varphi \in \mathscr{S}_0$, let ω be the function on G defined by the character of $Z(\mathbb{C}G)$ corresponding to the exceptional character associated with φ: then

$$\omega(x) = \frac{|G|}{|C_G(x)|} \cdot \frac{\varphi^\tau(x)}{\varphi^\tau(1)}. \tag{3.2}$$

We note that ω is constant on $H_m^\#$ since $H_m \subseteq \ker\varphi$, by Corollary 19. Now multiplication in $Z(\mathbb{C}G)$ implies the following relations between characters and class algebra constants, as in the proof of Corollary 2.9.

Lemma 20. *Let $g, h \in H_m^\#$. Then*

$$\varphi^\tau(1)\omega(g)\omega(h) = \varphi^\tau(1)\sum_x a_{ghx}^G \omega(x)$$

where the summation is taken over representatives of the conjugacy classes of G.

We now aim to replace these relations by congruences modulo $|H|$ in the ring of integers in an algebraic number field containing all character values for G. Since elements of H which are conjugate in G are already conjugate in N, if $g, h \in H_m^{\#}$ and $x \in H$ we can associate with each class algebra constant a_{ghx}^G of $Z(\mathbb{C}G)$ a unique class algebra constant a_{ghx}^N of $Z(\mathbb{C}G)$. We note that, as in Proposition 2.51,

$$a_{ghx}^G = |\mathscr{A}_G(g, h; x)|$$

where

$$\mathscr{A}_G(g, h; x) = \{(g', h') \mid g' \sim_G g, h' \sim_G h \text{ and } g'h' = x\};$$

$a_{ghx}^N = |\mathscr{A}_N(g, h; x)|$ with an analogous definition. (Here, by $g' \sim_G g$, we mean that g and g' are conjugate in G: the suffix will be omitted since, in our context, elements of H which are conjugate in G are already conjugate in N.)

Lemma 21. *Let $g, h \in H_m^{\#}$ and let $x \in H$. Then the following hold.*

(i) $a_{ghx}^G \equiv a_{ghx}^N \pmod{|C_H(x)|}$.

(ii) $a_{ghx}^N = 0$ *if* $x \notin H_m$.

(iii) $a_{gh1}^N = \begin{cases} e & \text{if } h \sim g^{-1}, \\ 0 & \text{otherwise.} \end{cases}$

(iv) $\sum_x a_{ggx}^N = e$ *and* $\sum_x a_{gg^{-1}x}^N = e - 1$, *where the summation is taken over representatives of the N-conjugacy classes in $H_m^{\#}$.*

Proof. $C_H(x)$ acts on $\mathscr{A}_G(g, h; x)$ by conjugation. If (g', h') does not lie in an orbit of length $|C_H(x)|$, then

$$\langle g', h' \rangle \subseteq C_G(y) \subseteq H$$

for some element $y \in C_H(x)^{\#}$, so that $(g', h') \in \mathscr{A}_N(g, h; x)$. So (i) holds. Now, for $(g', h') \in \mathscr{A}_N(g, h; x)$, we see that $g'h' \in H_m$ so that (ii) holds.

Let g^N denote the conjugacy class of g in N. Since e is odd, $g \not\sim g^{-1}$. Since $g \in Z(H)$, $|g^N| = e$ and (iii) is clear: to see (iv) we count elements in the product $C_g C_h$ (computed in $Z(\mathbb{C}N)$) to get

$$e^2 = |g^N|^2 = \sum_x a_{ggx}^N |x^N| = e \sum_x a_{ggx}^N$$

and

$$e^2 = |g^N| \cdot |(g^{-1})^N| = \sum_{x \neq 1} a_{gg^{-1}x}^N |x^N| + a_{gg^{-1}1}^N.$$

Corollary 22. *The following congruences modulo $|H|$ hold for $g \in H_m^{\#}$:*

$$\varphi^{\tau}(1)\omega(g)^2 \equiv e\varphi^{\tau}(1)\omega(g),$$

$$\varphi^{\tau}(1)\omega(g)^2 \equiv (e - 1)\varphi^{\tau}(1)\omega(g) + e\varphi^{\tau}(1).$$

Proof. If x is not conjugate to any element of H, then from (3.2)

$$\varphi^\tau(1)\omega(x) \equiv 0 \pmod{|H|},$$

while, for $x \in H - H_m$,

$$\varphi^\tau(1)\omega(x) \equiv 0 \pmod{[H:C_H(x)]}.$$

Now (i) and (ii) allow us to restrict the summation in Lemma 20 to N-conjugacy classes of H_m if we replace equality by a congruence modulo $|H|$. Since $H_m \subseteq \ker \varphi$, Corollary 19 implies in particular that φ^τ is constant on $H_m^\#$ as then is ω. Thus we may substitute the values given by (iii) and (iv) into this congruence, first with $g = h$ and then with $g^{-1} = h$, to obtain the two congruences claimed.

Corollary 23. *f_m divides c_m.*

Proof. Let $g \in H_m^\#$. Corollary 22 implies the congruence

$$\varphi^\tau(1)\omega(g) \equiv e\varphi^\tau(1) \pmod{|H|}.$$

Substituting the value for $\omega(g)$ given by (3.2), we obtain a congruence

$$[G:H]\varphi^\tau(g) \equiv e\varphi^\tau(1) \pmod{|H|}.$$

Then $\varphi^\tau(g) \equiv \varphi^\tau(1) \pmod{|H|}$, and we may apply Corollary 19.

Proposition 24. *$\mathscr{S}_0 \cup \mathscr{S}_m$ is coherent.*

Proof. Let $\mathscr{S}_0 = \{\varphi_1, \ldots, \varphi_t\}$ and pick $\psi \in \mathscr{S}_m$ with $\psi(1) = ef_m$. By Lemma 4, it is sufficient to show that $\mathscr{S}_0 \cup \{\psi\}$ is coherent. Let $\mu = f_m\varphi_1 - \psi$. Then $\mu \in \mathbb{Z}_0(\mathscr{S})$ and

$$\|\mu^G\|_G^2 = \|\mu\|_N^2 = f_m^2 + 1. \tag{3.3}$$

For $i = 1, \ldots, t$, put

$$\varphi_i^* = \varphi_i + \sum_{k=0}^{m} c_k\alpha_k + d\cdot 1_N$$

where $c_0, \ldots, c_m, d$ are defined as in Lemma 18. Then, since μ vanishes on $N - H$,

$$(\mu^G, \varphi_i^\tau)_G = (\mu, \varphi_i^\tau|_N)_N = (\mu, \varphi_i^*)_N$$

for $1 \leqslant i \leqslant t$. By Corollary 23, we may put $c_m = cf_m$: then

$$(\mu^G, \varphi_i^\tau)_G = f_m(c_0 + \delta_{1i}) - f_m c.$$

Thus f_m divides $(\mu^G, \varphi_i^\tau)_G$: since $f_m > 1$, it follows from (3.3) that $(\mu^G, \varphi_i^\tau)_G \neq 0$ for exactly one value of i.

If $i \geqslant 2$, then

$$(\mu^G, \varphi_1^\tau - \varphi_i^\tau)_G = (\mu, \varphi_1 - \varphi_i)_N = f_m.$$

If $t \geqslant 3$, then this holds for two different values of i and hence

$$(\mu^G, \varphi_1^\tau)_G = f_m:$$

if $t = 2$, we may choose the extension of τ from $\mathbb{Z}_0(\mathscr{S}_0)$ to $\mathbb{Z}(\mathscr{S}_0)$ to guarantee this. Now there is an irreducible character χ of G, with $\chi \notin \pm \mathscr{S}_0^\tau$, such that $(\mu^G, \chi) = \pm 1$: we define $\psi^\tau = \pm \chi$, the sign being chosen so that

$$(f_m \varphi_1^\tau - \psi^\tau)(1) = 0.$$

Then $((\mathscr{S}_0 \cup \{\psi\}), \tau)$ is coherent.

This completes the proof of Theorem 14.

We now obtain an improvement of Theorem 13. In contrast to the proof of Theorem 14, the delicate work takes place when the T.I.-set is an abelian p-group and a detailed character theoretic analysis of the class algebra constants is needed.

Theorem 25. *Let G be a group satisfying Hypothesis III, and assume that $\mathscr{S}$ is coherent. Assume, further, that*

(i) $e < \frac{1}{2}(|H| - 1)$, *and*
(ii) *if H is a nonabelian 2-group, then $(e+1)^2 \leqslant |H|$.*

Then, for $\varphi \in \mathscr{S}$,

$$\varphi^\tau(g) = \varphi(g)$$

for all $g \in H^\#$.

If H is an abelian p-group for p an odd prime, then this result is due to Sibley (1975), and our first task will be to reduce to that situation. After Theorem 14, the assumption of coherence is relevant only if H is a nonabelian 2-group with $(e+1)^2 \leqslant |H| \leqslant 4e^2 + 1$, although the proof is independent of that theorem. Condition (i) is equivalent to the assertion that $H^\#$ contains at least three N-conjugacy classes: in this form, it is vital to the proof. Of course, one cannot expect *precisely* the same result if $e = \frac{1}{2}(|H| - 1)$ since, in this case, H is certainly an elementary abelian p-group for some odd prime p so that two possibilities for the extension of τ to $\mathscr{S}$ will exist, but one might reasonably hope that there might be an inequality $|c| \leqslant 1$ to allow for this indeterminacy. If H is cyclic (and therefore of prime order), then deep results of Brauer (1942) using modular character theory imply this; otherwise, no direct proof is known, although it is not difficult to verify this conclusion using the classification of finite simple groups.

The proof of Theorem 25 is heavily computational: we shall simply number formulae as we go for later reference.

Let $\mathscr{S}=\{\varphi_1,\dots,\varphi_t\}$ and let $\chi_1,\dots,\chi_s$ be the nonexceptional characters of G. By Theorem 13, the nonexceptional characters are rational-valued on H. Since H is a T.I.-set, we have

$$\sum_{\beta=1}^{s}\chi_\beta(h)^2+\sum_{\alpha=1}^{t}|\varphi_\alpha^\tau(h)|^2=|C_H(h)| \tag{3.4}$$

for elements $h\in H^\#$. By Theorem 13, there exists a rational integer c such that

$$\varphi_\alpha^\tau(h)=\varphi_\alpha(h)+\frac{\varphi_\alpha(1)}{e}\cdot c \tag{3.5}$$

for $\alpha=1,\dots,t$: thus

$$\sum_{\alpha=1}^{t}|\varphi_\alpha^\tau(h)|^2=\sum_{\alpha=1}^{t}|\varphi_\alpha(h)|^2+\frac{c}{e}\sum_{\alpha=1}^{t}(\varphi_\alpha(1)\varphi_\alpha(h)+\varphi_\alpha(1)\overline{\varphi_\alpha(h)})+\frac{c^2}{e^2}\sum_{\alpha=1}^{t}\varphi_\alpha(1)^2.$$

By consideration of the orthogonality relations for the characters of N, and substituting the result in (3.4), this implies that

$$\sum_{\beta=1}^{s}\chi_\beta(h)^2=e+2c-\frac{c^2}{e}(|H|-1)\geqslant 1 \tag{3.6}$$

since no $\varphi_\alpha^\tau=1_G$.

Suppose first that $|H|\geqslant(e+1)^2$. Then the inequality (3.6) implies that

$$e+2c-c^2(e+2)>0,$$

and hence that $|c|<1$. So $c=0$ in this case.

If $|H|<(e+1)^2$, then H has only one N-chief factor, and so H is elementary abelian. If, furthermore, H were an elementary abelian 2-group, then the argument which proved Lemma 2.57 would show that H was a normal subgroup of G and there would be nothing to prove. So we now assume only that H is an abelian p-subgroup of order q for some odd prime p, and that $e<\frac{1}{2}(q-1)$. Then we have, with our given notation, $t=(q-1)/e$ and $t\geqslant 3$.

The character value formula (3.5) now becomes

$$\varphi_\alpha^\tau(h)=\varphi_\alpha(h)+c \tag{3.5$'$}$$

for $\alpha=1,\dots,t$: then, since $(1_H,\varphi_\alpha^\tau|_H)\in\mathbb{Z}$, we obtain a congruence

$$\varphi_\alpha^\tau(1)\equiv e+c \pmod q. \tag{3.7}$$

For the remainder of the proof, we assume that $c\neq 0$ and, without loss since Theorem 25 is certainly true if G has a normal p-complement (when H must be cyclic), that $e\geqslant 2$. Then the inequality (3.6) becomes

$$tc^2-2c-e<0:$$

since $t\geqslant 3$ and $3e-2\sqrt{e}-e=2\sqrt{e}\,(\sqrt{e}-1)>0$, this implies that

$$c^2<e. \tag{3.8}$$

(Note that, in fact, the same is true if $t = 2$ and $e \geqslant 3$: cf. Section 3.4 if $e \leqslant 4$.)

Let $\mathscr{C}_1, \ldots, \mathscr{C}_t$ denote the conjugacy classes of nonidentity p-elements and $C_1, \ldots, C_t$ the corresponding class sums, taken either in G or in N, according to the context which will always be clear. For indices i, j, k, let a_{ijk} and a^G_{ijk} denote the appropriate class algebra constants of $Z(\mathbb{C}N)$ and $Z(\mathbb{C}G)$ respectively. These indices, as also l, m, n, will be arbitrary in the range $1, \ldots, t$ unless specified. Also, let $g_i \in \mathscr{C}_i \cap N$.

Noting that e^2/q is an 'average' value for a_{ijk}, we define l_{ijk} by

$$a_{ijk} = \frac{e}{q}\left(e + \frac{1}{e} l_{ijk}\right). \tag{3.9}$$

Our goal is to exploit these differences. Applying Corollary 2.9, the character theoretic formula for a_{ijk}, we get the equation

$$a_{ijk} = \frac{qe}{q^2}\left(e + \frac{1}{e}\sum_{\alpha=1}^{t} \varphi_\alpha(g_i)\varphi_\alpha(g_j)\overline{\varphi_\alpha(g_k)}\right),$$

whence

$$l_{ijk} = \sum_{\alpha=1}^{t} \varphi_\alpha(g_i)\varphi_\alpha(g_j)\overline{\varphi_\alpha(g_k)}. \tag{3.10}$$

In G, we have, summing over all the irreducible characters χ of G,

$$a^G_{ijk} = \frac{|G|}{q^2}\sum \frac{\chi(g_i)\chi(g_j)\overline{\chi(g_k)}}{\chi(1)}:$$

by the same argument as in the proof of Lemma 21(i),

$$a^G_{ijk} \equiv a_{ijk} \pmod q,$$

so that

$$a_{ijk} \equiv \frac{|G|}{q^2}\sum_{\chi} \frac{\chi(g_i)\chi(g_j)\overline{\chi(g_k)}}{\chi(1)} \pmod q.$$

Since the nonexceptional characters of G are constant on $H^{\#}$ by Theorem 13, we may take the difference of the two class algebra constants to obtain the congruence

$$a_{ijk} - a_{lmn} \equiv \frac{|G|}{q^2 d}\sum_{\alpha=1}^{t} \left(\varphi^{\tau}_\alpha(g_i)\varphi^{\tau}_\alpha(g_j)\overline{\varphi^{\tau}_\alpha(g_k)} - \varphi^{\tau}_\alpha(g_l)\varphi^{\tau}_\alpha(g_m)\overline{\varphi^{\tau}_\alpha(g_n)}\right) \pmod q \tag{3.11}$$

where all the exceptional characters have the same degree d. The orthogonality relations, applied to the characters of N, give the formulae

$$\sum_{\alpha=1}^{t} \varphi_\alpha(g_i) = -1$$

and

$$\sum_{\alpha=1}^{t} \varphi_\alpha(g_i)\varphi_\alpha(g_j) = \delta_{ij'},$$

where $\mathscr{C}_{j'}$ is the conjugacy class of elements inverse to those in $\mathscr{C}_j$: expanding the right-hand side of (3.11) using the formulae (3.5′) and (3.10), the congruence (3.11) becomes

$$a_{ijk} - a_{lmn}$$
$$\equiv \frac{|G|}{q^2 d}(l_{ijk} - l_{lmn}) + \frac{|G|c}{qd}(\delta_{ij'} + \delta_{jk} + \delta_{ik} - \delta_{lm'} - \delta_{mn} - \delta_{ln}) \pmod q.$$

By (3.7) and (3.8), $d \not\equiv 0 \pmod q$: so, multiplying both sides of this congruence by d and applying the congruence (3.7), we obtain the further congruence

$$c(a_{ijk} - a_{lmn}) \equiv ce(\delta_{ij'} + \delta_{jk} + \delta_{ik} - \delta_{lm'} - \delta_{mn} - \delta_{ln}) \pmod q \qquad (3.12)$$

in view of the definition (3.9) and the congruence $[G:H] \equiv e \pmod q$.

We may now confine our calculations to the Frobenius group N in order to obtain a contradiction from the assumption that $c \neq 0$. Put $c = c_1 c_2$ where c_1 is a power of p and $(c_1, c_2) = 1$, and put $r = q/c_1$. Since, from (3.8), $c^2 < e$ and also $e < q$, we have $r^2 > q$ so that, in particular,

$$pq \text{ divides } r^2. \qquad (3.13)$$

In any case, the congruence (3.12) becomes

$$a_{ijk} - e(\delta_{ij'} + \delta_{jk} + \delta_{ik}) \equiv a_{lmn} - e(\delta_{lm'} + \delta_{mn} + \delta_{ln}) \pmod r: \qquad (3.14)$$

this congruence is independent of the particular indices, and we shall denote it by M. This is the critical step, from which the contradiction will be obtained. It will *fail* for some i, j, k, l, m, n: the remainder of the calculation is devised to avoid having to find (or compute) particular values.

As an exact analogue of Lemma 21(iv), we have equations

$$\sum_{k=1}^{t} a_{ijk} = e - \delta_{ij'}, \quad 1 \leqslant i, j \leqslant t:$$

summing the congruence (3.14) over k, we obtain a further congruence

$$e - \delta_{ij'} - e(t\delta_{ij'} + 2) \equiv Mt \pmod r.$$

Now, since $q = 1 + et$, this becomes

$$-e \equiv Mt \pmod r.$$

Hence $M \equiv e^2 \pmod r$ and so

$$a_{ijk} - e(\delta_{ij'} + \delta_{jk} + \delta_{ik}) \equiv e^2 \pmod r.$$

Define integers b_{ijk} by

$$a_{ijk} = e^2 + e(\delta_{ij'} + \delta_{jk} + \delta_{ik}) + rb_{ijk}. \qquad (3.15)$$

Since H is abelian, $g_i g_j = g_k$ if and only if $g_k^{-1} g_j = g_i^{-1}$, so that $a_{ijk} = a_{k'ji'}$ and $b_{ijk} = b_{k'ji'}$. Now, summing (3.15) over k, we obtain the relation

$$e - \delta_{ij'} = e^2 t + e(t + \delta_{ij'} + 2) + r \sum_{k=1}^{t} b_{ijk},$$

which reduces to

$$r\sum_{k=1}^{t} b_{ijk} = -q\delta_{ij'} - eq. \tag{3.16}$$

Let $\mathscr{C}_0 = \{1\}$. Then the analogue of Lemma 21(iii) gives the class algebra constant

$$a_{ij0} = e\delta_{ij'}.$$

Also

$$a_{0jk} = \delta_{jk}.$$

We now compute $\sum_{l=0}^{t} a_{ijl}a_{lkm}$. (Cf. Exercise 3 of Section 2.9.) Writing

$$\sum_{l=0}^{t} a_{ijl}a_{lkm} = e\delta_{ij'}\delta_{km} + \sum_{l=1}^{t} a_{ijl}a_{lkm}$$

and substituting in the sum on the right-hand side using (3.15), if we then multiply out and use the formula (3.16), a straightforward calculation yields that

$$\begin{aligned}\sum_{l=0}^{t} a_{ijl}a_{lkm} &= 3e^3 + e^2(\delta_{ij'} + \delta_{ik'} + \delta_{jk'} + \delta_{im} + \delta_{jm} + \delta_{km})\\ &\quad + er(b_{ikm} + b_{jkm} + b_{ijm} + b_{ijk'}) - e^3q\\ &\quad - e^2q\delta_{ij'} - e^2q\delta_{km} - eq\delta_{km}\delta_{ij'} + r^2\sum_{l=1}^{t} b_{ijl}b_{lkm}.\end{aligned} \tag{3.17}$$

Since $(C_iC_j)C_k = (C_jC_k)C_i$, taken as class sums in $\mathbb{C}N$,

$$\sum_{l=0}^{t} a_{ijl}a_{lkm} = \sum_{l=0}^{t} a_{jkl}a_{lim} \tag{3.18}$$

as each is the coefficient of C_m in the product. Since $t \geqslant 3$, we may take i, k, m distinct and take $j = i'$ in (3.18): evaluating both sides, using (3.17), we may obtain an equation

$$-e^2q + r^2\sum_{l=1}^{t} b_{ii'l}b_{lkm} = r^2\sum_{l=1}^{t} b_{i'kl}b_{lim}.$$

So r^2 divides e^2q. However, pq divides r^2 by (3.13): this implies that p divides e, which is not so. Theorem 25 is proved.

In view of the comment after the proof of (3.8), we note that it is only in this final step that we use the assumption that $t \geqslant 3$: it can be verified that no contradiction arises in this way when $t = 2$ (allowing for the fact that the best possible outcome is $|c| \leqslant 1$).

Theorem 25 thus provides a further situation in which there is a best possible answer to the problem posed in the last chapter, namely to discover when a portion of the character table can be lifted from that of a subgroup to that of the whole group.

4 CN-groups of odd order

We recall that a finite group is said to be a *CN-group* if the centraliser of every nonidentity element is nilpotent. As the first of our applications of Theorem 14, we shall give a proof of the Feit–Hall–Thompson theorem (1960). Curiously, although Theorem 25 would appear to be applicable, it provides no useful information.

Theorem 26. *There does not exist a nonabelian simple CN-group of odd order.*

Sibley's result enables us to follow the character theoretic arguments applied in the last chapter for CA-groups, though the group theoretic reductions are now highly nontrivial.

Throughout this section, we let G be a minimal counterexample: that is, a nonabelian simple CN-group of least possible odd order. We first examine the subgroup structure in order that we may then apply the techniques of exceptional characters.

Lemma 27. *Let H be a maximal nilpotent subgroup of G. Assume that $|H|$ is divisible by at least two distinct primes. Then H is a Hall subgroup of G. Furthermore, H is a T.I.-set in G and $N_G(H)$ is a Frobenius group with Frobenius kernel H.*

Proof. Let P be a Sylow p-subgroup of H. If P is properly contained in a Sylow p-subgroup P^* of G, let $x \in (P \cap Z(N_{P^*}(P)))^\#$; then $C_G(x) \supseteq \langle H, N_{P^*}(P) \rangle$, contrary to the maximality of H. So H is a Hall subgroup of G. Now $H = P \cdot C(P)$, since $P \cdot C(P)$ is certainly nilpotent and contains H. In particular, it follows that H is the unique maximal nilpotent subgroup of G containing P, and $N(P) = N(H)$.

Let Q be a Sylow q-subgroup of H with $q \neq p$. If $x \in H^\#$, we have $Q \subseteq C(x)$: by the above remarks, $C(x) \subseteq H$. If $y \in P^*$ and $x = y^g$ for some $g \in G$, then Q and Q^g are Sylow subgroups of $C(x)$ so that $Q = Q^g$ and $g \in N(H)$. Thus P is a T.I.-set, and so then is H. Now, if $N(H) = H$, it follows from the focal subgroup theorem (Theorem 1.46) that

$$P \cap G' = \langle xy^{-1} | x, y \in P, y = x^g \text{ for some } g \in G \rangle = \langle [x^{-1}, g] | x, g \in P \rangle = P',$$

contrary to the simplicity of G. So $N(H) \neq H$, and $N(H)$ is a Frobenius group since H contains the centralisers of each of its nonidentity elements of prime power order.

The proof just given displays vividly the power that can be achieved by elementary arguments when there is an interplay between two primes.

When there are isolated primes, however, far deeper arguments are needed and, in particular, we shall apply Thompson's normal p-complement theorem (stated as Theorem 1.25).

It will be useful to introduce some standard group theoretic notation. Let G be a finite group and let p be a prime. Then $O_p(G)$ denotes the largest normal p-subgroup of G and $O_{p'}(G)$ denotes the largest normal subgroup of G of order coprime to p. The subgroups $O_{p',p}(G)$ and $O_{p,p'}(G)$ are the subgroups containing $O_{p'}(G)$ and $O_p(G)$, respectively, such that

$$O_{p',p}(G)/O_{p'}(G) = O_p(G/O_{p'}(G))$$

and

$$O_{p,p'}(G)/O_p(G) = O_{p'}(G/O_p(G)).$$

Lemma 28. *Let P be a Sylow p-subgroup of G which is a maximal nilpotent subgroup of G. Then P is a T.I.-set in G and $N(P)$ is a Frobenius group.*

Proof. In view of Lemma 27, G satisfies the following condition:

(CN_p) G is a CN-group in which the centraliser of every nonidentity p-element is a p-group.

We observe that this condition may be verified by reference only to nonidentity elements of order coprime to p and then that the property is inherited in quotients by normal p-subgroups.

If P_0 is any nonidentity p-subgroup of G and $N = N_G(P_0)$, then $N/O_p(N)$ is a CN_p-group. Suppose that N is not a p-group. Since G is a minimal counterexample to Theorem 26, $N/O_p(N)$ has no nonabelian simple subgroups. Let L be a subgroup of N containing $O_p(N)$ such that $L/O_p(N)$ is a normal elementary abelian q-subgroup of $N/O_p(N)$ for some prime $q \neq p$. Let Q be a Sylow q-subgroup of L. Since $C_{O_p(N)}(x) = 1$ for all $x \in Q^\#$, it follows that L is a Frobenius group with Q as a cyclic Frobenius complement. (See Exercise 3 of Section 2.4.) By a Frattini argument, $N = L \cdot N_N(Q)$: thus N is soluble. Indeed, since $C_N(Q)$ is a nilpotent group of order prime to p and $N(Q)/C(Q)$ is cyclic, either $N/O_p(N)$ is nilpotent or $N/O_p(N)$ is a Frobenius group whose Frobenius complement is cyclic and, since L is not nilpotent, that complement is a p-group. In either case, $O_{p,p'}(N)$ is a Frobenius group with a cyclic Frobenius complement. In the first case, if $P_0 = O_p(N)$, then P_0 is a Sylow p-subgroup of G. In the latter case, N is a particular example of what is called a *three-step group*, and we observe that if P_1 is a p-subgroup of N not contained in $O_p(N)$, then $N_N(P_1)$ is a p-group.

Suppose that P is not a T.I.-set in G. Amongst other conjugates of P, pick $\tilde{P}$ with $U = P \cap \tilde{P}$ maximal. Let M be a subgroup of G containing

$N_G(U)$, maximal subject to the condition that $O_p(M) \neq 1$. Since $N_P(U) \cap N_{\tilde{P}}(U) = U$, we see that $N_G(U)$ is a three-step group: hence, so is M. Let P^* be a Sylow p-subgroup of M. Then $C_M(Z(P^*)) = P^*$ so that, since M does not have a normal p-complement, nor does $N_M(J(P^*))$ in view of Thompson's normal p-complement theorem (Theorem 1.25). Thus, by the remark at the end of the paragraph above, $J(P^*) \subseteq O_p(M)$. Now $J(P^*) = J(O_p(M))$: hence $M = N_G(J(P^*))$ by the maximality of M, so that P^* is a Sylow p-subgroup of G. Indeed, we may suppose that $P^* \supseteq N_P(U)$ so that, by the choice of U, we have $P^* = P$. Also, since $N_G(U)$ is not a p-group, $U \subseteq O_p(M)$.

Thus we have shown that, if P is not a T.I.-set and $M = N_G(J(P))$, then $P \cap \tilde{P} \subseteq O_p(M)$ *whenever* $\tilde{P}$ is a Sylow p-subgroup of G distinct from P. Also, $P = N_G(P)$ since $N_G(P) \subseteq N_G(J(P)) = M$ and $P \nsubseteq O_p(M)$. Let $x \in P - O_p(M)$ and suppose that $x^g \in P$ for some element $g \in G$. Then $x \in P \cap P^{g^{-1}}$ so that $P = P^{g^{-1}}$ and $g \in P$. By the focal subgroup theorem (Theorem 1.46),

$$P \cap G' = \langle ab^{-1} \mid a, b \in P, b = a^g \text{ for some } g \in G \rangle \subseteq O_p(M),$$

contrary to the simplicity of G. So P is a T.I.-set. If $P = N_G(P)$, then the focal subgroup theorem implies that $P \cap G' = P'$, again contrary to the simplicity of G. Since G is a CN_p-group, $N_G(P)$ must be a Frobenius group.

The fact that the maximal nilpotent subgroups are Hall subgroups, and that any Sylow subgroup lies in a unique such Hall subgroup, enables us to partition the prime divisors of $|G|$, and we need only show the nonexistence of a group satisfying the following conditions.

Hypothesis IV. G is a nonabelian simple group of odd order. The prime divisors of $|G|$ fall into disjoint subsets $\pi_1, \ldots, \pi_s$. For $i = 1, \ldots, s$, there is a nilpotent Hall π_i-subgroup H_i of order h_i such that

(a) H_i is a T.I.-set in G, and
(b) if $N_i = N_G(H_i)$, then N_i is a Frobenius group of order $h_i e_i$, whose Frobenius kernel is H_i.

We now assume that G is a group satisfying this hypothesis. The proof of nonexistence follows the arguments given for CA-groups in the previous chapter, and we will only highlight the few differences, leaving the details as an exercise.

First we observe that, if H_i is nonabelian, then the number of characters of N_i which do not have H_i in their kernels is less than $(|H_i| - 1)/e_i$. For each $i = 1, \ldots, s$, we let r_i be this number, which is also the number of

G-conjugacy classes of elements in $H_i^\#$ since H_i is a T.I.-set. Let $\mathscr{S}_i = \{\varphi_{i1}, \ldots, \varphi_{ir_i}\}$ be the set of irreducible characters of N_i which do not have H_i in their kernels. We may suppose that $\varphi_{i1}(1) = e_i$, though not all degrees are the same. Also, none of these characters is real.

By Theorem 14, $\mathscr{S}_i$ is coherent for each i and, if we let τ (generically) denote the corresponding isometry from $\mathbb{Z}_0(\mathscr{S}_i)$ to $\mathscr{C}\mathscr{h}(G)$, we may apply Theorem 13 where previously, in Chapter 3, we had applied Theorem 3.16. In particular, the rationality of the exceptional characters φ_{ij}^τ on elements not conjugate to elements of $H_i^\#$ and the constancy of nonexceptional characters on $H_i^\#$ enable us to deduce the following by counting characters and conjugacy classes.

Lemma 29. *Every nonprincipal irreducible character of G is an exceptional character associated with exactly one of the sets $\mathscr{S}_1, \ldots, \mathscr{S}_s$.*

In contrast with the situation for CA-groups, the characters in $\mathscr{S}_i^\tau$ do not take the same values on $H_k^\#$ for $k \neq i$: however, they *are* either all zero or all nonzero.

Lemma 30. *If g is an element not conjugate to an element of $H_i^\#$, then, for each j,*

$$\varphi_{ij}^\tau(g) = \frac{\varphi_{ij}(1)}{e_i}\varphi_{i1}^\tau(g),$$

and this value is constant and rational on $H_i^\#$.

Proof. This is immediate from the observation that

$$(\varphi_{i1}(1)\varphi_{ij} - \varphi_{ij}(1)\varphi_{i1}) \in \mathbb{Z}_0(\mathscr{S}_i),$$

together with Theorem 13(iv).

This enables us to partition the character table for G (see Table 4.2) in exactly the same way as we did for CA-groups to get Table 3.2, except that Theorem 13 tells us that the entries in a block A_{ii} are of the form

$$\varphi_{ij}^\tau(h_i) = \varphi_{ij}(h_i) + \frac{\varphi_{ij}(1)}{e_i}c_i,$$

while, by Lemma 30, the rows of each off-diagonal block A_{ik} are constant and, in particular, either all zero or all nonzero. Note that Theorem 25 would give $c_i = 0$ provided that $H_i^\#$ contains three or more G-conjugacy classes, but this need not necessarily be so.

We may now prove Theorem 26 by following precisely the same arguments

Table 4.2

	1	$\leftarrow H_1^\# \rightarrow$		$\leftarrow H_s^\# \rightarrow$
1	1	$1 \cdots 1$	$\cdots$	$1 \cdots 1$
$\varphi^\tau_{1j}, 1 \leqslant j \leqslant t_1$ {		A_{11}		A_{1s}
			$\ddots$	
$\varphi^\tau_{sj}, 1 \leqslant j \leqslant t_s$ {		A_{s1}		A_{ss}

as for Theorem 3.20, using Theorem 13 in place of Theorem 3.16 for information about character values.

The original proof of this result differed in two major ways. Thompson's normal p-complement theorem was not available in the concise form that it has been applied here, so that the group theoretic reductions, while having the same goal, were more complex, and the character theory was handled in a different way in the absence of Sibley's theorem. But that proof was a prototype for Feit and Thompson's proof that every group of odd order is soluble. In the reductions there, one must get precise information about the structure of maximal subgroups, and three-step groups can occur. For this reason, the character theoretic methods that we have so far described are inadequate, and the concept of special classes has to be widened to that of tamely imbedded subsets: also, mere induction of characters is insufficient, and better isometries must be constructed. It will be beyond the scope of this book to go into these ideas in detail or to describe Sibley's work on revising the calculations. However, in the final chapter we shall discuss other isometries and describe, in particular, an isometry due to Dade (1964) which picks out the essence of the method of tamely imbedded subsets and which has been used in the simplifications to which we have referred.

Exercise

1. Fill in the details of the character theory needed to prove Theorem 26.

5 Zassenhaus groups

A finite group which acts faithfully as a group of permutations on a set Ω is called a *Zassenhaus group* if the following three conditions are satisfied:

(i) G acts doubly transitively on Ω,
(ii) no nonidentity element of G fixes three or more points of Ω, and
(iii) G does not possess a regular normal subgroup (that is, a transitive subgroup of order $|G|$ in which nonidentity elements are fixed-point-free).

Condition (iii) ensures, in view of Frobenius' theorem, that some nonidentity element fixes two points. The study of Zassenhaus groups is thus a next stage after that of Frobenius groups. The first known examples were the groups $\mathrm{PSL}(2,q)$ (together with certain subgroups of their automorphism groups) in their natural action on the $q+1$ points of the projective line. Zassenhaus (1936) attempted to show that these were the only such groups, but the Suzuki groups which were discussed in Section 3 also satisfy these conditions, and they were discovered in the process of classifying all Zassenhaus groups: in fact, Suzuki's work together with that of others shows that these groups provide the complete answer.

Let G be a Zassenhaus group, and let G_0 be the stabiliser of a point. Then G_0 is a Frobenius group. Suppose that $|\Omega| = h+1$. Then $|G| = (h+1)he$, where e is the order of a two-point stabiliser, and $e > 1$ since G is itself not a Frobenius group. If $e = h-1$, then G is triply transitive and Zassenhaus' first result (Proposition 2.45) showed that such a group necessarily contained $\mathrm{PSL}(2,q)$ as a subgroup of index 1 or 2, acting on the projective line. He also obtained the same conclusion under the assumption that G is doubly transitive and e is even, in which case an elementary argument shows that $e \geqslant \frac{1}{2}(h-1)$, and his arguments can be extended to the case where $e = \frac{1}{2}(h-1)$ and e is odd. Zassenhaus' work represented the first characterisation of a family of finite simple groups and it remains of fundamental importance, first since subsequent characterisation of the groups $\mathrm{PSL}(2,q)$ always fall back on to his theorem for the identification (as, for example, Theorems 2.44 and 2.56), and secondly as a model for the identification of other groups from permutation theoretic properties. However, we shall not consider his work here, but refer the reader to Huppert and Blackburn [HB III; Ch. XI] where a complete classification of Zassenhaus groups is given.

In this section, we shall consider a major reduction step in that classification where Theorem 14 will permit a unified treatment. To fix notation, let G be a Zassenhaus group acting on a set Ω, where $|\Omega| = h+1$ and $|G| = (h+1)he$. Let $0, 1 \in \Omega$ and let the point stabiliser $G_0 = HE$, where H is the Frobenius kernel of G_0 and E is the two-point stabiliser G_{01}.

Theorem 31. *Let G be a Zassenhaus group, as above, and assume that e is odd. Then either*

(i) $e \geqslant \frac{1}{2}(h-1)$, *or*
(ii) *H is a nonabelian 2-group with* $[H:H'] \leqslant 4e^2+1$.

We remark that, in case (i), H must necessarily be elementary abelian, so that $h=p^n$ for some prime p. Then, by the earlier discussion, G must contain $\mathrm{PSL}(2,q)$ as a subgroup of index 1 or 2.

A theorem of this type was first proved by Feit (1960), and his version of Theorem 6 was developed for this purpose, having as the exceptional case (ii) only that H be a p-group. Itô (1962) then eliminated the possibility that p was odd, using modular character theory, and his result was later reproved by Glauberman (1969) using a detailed argument which did not, however, require modular characters. In the case $p=2$, Suzuki discovered his groups and showed that they completed the classification.

We shall now prove Theorem 31. The main steps will be the reduction to the case where G is a simple group, and H and E are T.I.-sets whose normalisers are Frobenius groups. Then we can apply the methods of exceptional characters, noting that the second possible conclusion is precisely the situation where coherence cannot be assured for characters of $N(H)$.

Let G be a Zassenhaus group for which, in the notation given, e is odd, and assume that G is not triply transitive; that is, $e<h-1$.

Lemma 32. *H is a T.I.-set, with normaliser HE.*

Proof. Since each nonidentity element of H fixes only the point 0, if $g \in G - G_0$, then H^g fixes $0g$ and $H \cap H^g = 1$. Now G_0 is a maximal subgroup of G since G is doubly transitive, so that $G_0 = N_G(H)$ and H is a T.I.-set.

Lemma 33. *If N is a nonidentity normal subgroup of G, then N is a Zassenhaus group of degree* $h+1$.

Proof. Since G is doubly transitive, N must be transitive. We claim that $N \cap E \neq 1$. For, if $N \cap E = 1$, then N is a Frobenius group whose Frobenius kernel must be a regular normal subgroup of G, and no such subgroup exists. Now, if $E_1 = N \cap E$, we have, by Proposition 2.41,

$$H \subseteq \langle E_1^H \rangle \subseteq \langle E_1^G \rangle \subseteq N:$$

thus $H \subseteq N_0$ where N_0 is the stabiliser in N of the point 0, so that N is doubly transitive on Ω. If N contained more than one regular normal subgroup, two of them would generate their direct product since G is doubly transitive, which is impossible since $(h+1)^2$ does not divide $|G|$:

thus any regular normal subgroup of N is also a regular normal subgroup of G, which has none. So N is a Zassenhaus group.

Corollary 34. *In order to prove Theorem* 31, *we may assume that G is simple.*

Proof. If not, let N be a minimal normal subgroup of G. Then N satisfies the hypotheses of Theorem 31 with $e_1 = |N \cap E|$ in place of e. So, if conclusion (ii) fails for G, it fails for N also and, if $e_1 \geqslant \frac{1}{2}(h-1)$, then $e \geqslant \frac{1}{2}(h-1)$. So we may replace G by N, and then again by a minimal normal subgroup of N, which will be simple.

We make this assumption for the remainder of the proof, noting that it is possible to show that a Zassenhaus group of odd degree is necessarily simple. (Exercise 1.)

Lemma 35. *E is a T.I.-set whose normaliser is a dihedral group of order* $2e$.

Proof. E is the 2-point stabiliser G_{01}. If $E \neq E^g$, then any element of $E \cap E^g$ fixes at least two points. Hence E is a T.I.-set. Now $N_G(E)$ stabilises the unordered pair $\{0, 1\}$: then $g^2 \in E$ whenever $g \in N_G(E)$. As G is doubly transitive, there exists an involution τ which interchanges 0 and 1, and $N_G(E) = E\langle\tau\rangle$.

Since e is odd, e is coprime to both h and $h+1$ so that E is a Hall subgroup of G and, by Proposition 2.41, E is metacyclic with cylic Sylow subgroups. We shall show that E is abelian. Let p be a prime divisor of the index $[E:E']$ and let P be a τ-invariant Sylow p-subgroup of E. Then either τ inverts P or τ centralises P. If the latter, then $P \subseteq Z(N_E(P)) = Z(N_G(P))$ since E is a T.I.-set, and thus G has a normal p-complement by Burnside's transfer theorem, which is not so. Thus $P\langle\tau\rangle$ is a dihedral group. Now let Q be any Sylow subgroup of E'. Then $P\langle\tau\rangle$ normalises Q and hence P centralises Q since Q is cyclic. But then $E' \subseteq Z(E)$ and so E is nilpotent; since E has cyclic Sylow subgroups, it follows that E is cyclic and that $N_G(E)$ is dihedral.

Lemma 36. *G contains at least two conjugacy classes of elements which act fixed-point-freely on* Ω.

Proof. Since $|G_0| = he$ and, as e is odd, $h+1$ is coprime to both h and e, any nonidentity element of order dividing $h+1$ acts fixed-point-freely on Ω.

Suppose that all such elements are conjugate. Then, in particular, they have the same prime order, p say, and $h+1=p^n$ for some positive integer n. Let P be a Sylow p-subgroup of G. If $p=2$, then P is elementary abelian. If p is odd, then $|N(P)|$ is even since all elements of $Z(P)$ are conjugate in $N(P)$: but now the centraliser of an involution has order dividing h, so that $N(P)$ contains an involution which acts fixed-point-freely on P and, again, P is elementary abelian.

In either case, since H and E are T.I.-sets and, in particular, contain the centralisers of each of their nonidentity elements, $P=C_G(x)$ for any $x\in P^{\#}$, and $|N(P)|=p^n(p^n-1)=(h+1)h$ since all nonidentity elements of P are assumed conjugate. Furthermore, P is also a T.I.-set so that P has at least $(1+p^n)$ conjugates. So $|G|\geqslant(h+2)(h+1)h$, which is not the case. So G must contain at least two conjugacy classes of fixed-point-free elements.

We are now in a position to apply character theory to complete the proof of Theorem 31. Since G is doubly transitive, G has an irreducible character χ such that $1+\chi$ is the character of the permutation representation, and

$$\chi(g)=\begin{cases} h & \text{if } g=1,\\ 0 & \text{if } g \text{ is conjugate to an element of } H^{\#},\\ 1 & \text{if } g \text{ is conjugate to an element of } E^{\#},\\ -1 & \text{otherwise.}\end{cases}$$

We note, in particular, that $h\equiv 1 \pmod e$. Since E is strongly self-centralising and $[N(E):E]=2$, we may apply Theorem 3.17. Let $t=\frac{1}{2}(e-1)$ and let $\psi_1,\dots,\psi_t$ be the nonlinear characters of $N(E)$. Then G has t irreducible characters $\Psi_1,\dots,\Psi_t$ (exceptional if $t\geqslant 2$) such that the character χ above is as in Theorem 3.17 with $\varepsilon=1$ and

$$\Psi_i(g)=\begin{cases} h+1 & \text{if } g=1,\\ 1 & \text{if } g \text{ is conjugate to an element of } H^{\#},\\ \psi_i(g) & \text{if } g\in E^{\#},\\ 0 & \text{if } g \text{ is fixed-point-free on } \Omega,\end{cases}$$

for each $i=1,\dots,t$. Furthermore, if $\lambda_1,\dots,\lambda_t$ are characters of E such that $\psi_i=\lambda_i^{N(E)}$ for $i=1,\dots,t$, we may observe by direct computation that, if $\tilde{\lambda}_1,\dots,\tilde{\lambda}_t$ are the characters of G_0 having H in their kernels and $\tilde{\lambda}_i|_E=\lambda_i$, then $\Psi_i=\tilde{\lambda}_i^G$ for each i. (Compare this with the construction in Section 2.7 of the character table of $\mathrm{SL}(2,2^n)$ using Mackey's decomposition theorem.)

Now suppose that conclusion (ii) of Theorem 31 does not hold. Let $\varphi_1,\dots,\varphi_r$ be the irreducible characters of G_0 which do not have H in their kernels, where r is the number of nonidentity G-conjugacy classes in H and, by assumption, $r\geqslant 2$. Then $\{\varphi_1,\dots,\varphi_r\}$ is coherent by Theorem 14 and, in particular, G has a set of r exceptional characters $\Phi_1, \ldots, \Phi_r$ such that

$$(\varphi_j(1)\Phi_i - \varphi_i(1)\Phi_j) = \pm(\varphi_j(1)\varphi_i - \varphi_i(1)\varphi_j)^G$$

for $1 \leqslant i,j \leqslant r$. These are, in fact, necessarily distinct from the characters already obtained, but we shall not need this information. (See Exercise 3.)

We have constructed $r+t+2$ characters but, by Lemma 36, G contains at least $r+t+3$ conjugacy classes. Let ξ be an irreducible character distinct from those already obtained. We shall compute $\xi(1)$ by considering $\xi|_{G_0}$. Applying Frobenius' reciprocity theorem, we see that

$$(\xi|_{G_0}, 1_{G_0}) = (\xi, 1_G + \chi) = 0,$$
$$(\xi|_{G_0}, \tilde{\lambda}_i) = (\xi, \Psi_i) = 0$$

for $i = 1, \ldots, t$, and that

$$(\xi|_{G_0}, (\varphi_j(1)\varphi_i - \varphi_i(1)\varphi_j))_{G_0} = \pm(\xi, (\varphi_j(1)\Phi_i - \varphi_i(1)\Phi_j))_G = 0$$

for $1 \leqslant i,j \leqslant r$. Thus, since $1_{G_0}, \tilde{\lambda}_1, \ldots, \tilde{\lambda}_t, \varphi_1, \ldots, \varphi_r$ are all the characters of G_0 and $\xi|_{G_0} \not\equiv 0$, there is an integer κ such that

$$\xi|_{G_0} = \kappa \sum_{i=1}^{r} \frac{\varphi_i(1)}{e} \varphi_i$$

where $\varphi_1(1) = e$. By the orthogonality relations applied to G_0,

$$\sum_{i=1}^{r} \varphi_i(1)^2 = he - e:$$

hence $h-1$ divides $\xi(1)$.

Now $h-1$ divides $|G|$ and, since $|G| = (h+1)he$, it follows that $\frac{1}{2}(h-1)$ divides e. This completes the proof of Theorem 31.

Exercises

1. Let G be a Zassenhaus group of odd degree $h+1$. Let H and E be as defined at the beginning of the section, and suppose that G is not simple. Let N be a minimal normal subgroup and put $F = N \cap E$.

   (i) Show that N is a simple Zassenhaus group of degree $h+1$ having odd index in G.

   (ii) Show that $N_N(F)$ and $N_G(E)$ are dihedral groups.

   (iii) Show that $N_N(F)$ is a proper normal subgroup of odd index in $N_G(E)$. Deduce that, in fact, G must be simple.

   What can be said about normal subgroups of Zassenhaus groups of even degree?

2. Let G be a Suzuki group $\mathrm{Sz}(q)$ where $q = 2^n$ considered as a Zassenhaus group of degree $q+1$. Suppose that we know that G has subgroups

A_1 and A_2 which are strongly self-centralising subgroups of index 4 in their normalisers.

(i) Show that A_1 and A_2 share nonexceptional characters.

(ii) By considering the exceptional characters that arise from $N_G(E)$ and the character of degree q arising from the permutation representation (which was not available to us in Section 2), show that G has exactly two conjugacy classes of elements of order 4.

3. Continue the analysis of characters in the proof of Theorem 31 to obtain the following additional information.

(i) Prove that the characters $\Phi_1, \ldots, \Phi_r$ are distinct from the characters $\chi, \Psi_1, \ldots, \Psi_t$.

(ii) Assume that $e < \frac{1}{2}(h-1)$. Use Theorem 25 to show that

$$\Phi_i(g) = \varphi_i(g) \quad \text{for } g \in H^\#$$
$$\Phi_i(1) = \varphi_i(1).$$

[Note. Theorem 25 shows that '$c = 0$'. Then obtain the congruence $\Phi_1(1) \equiv \pm e \pmod{h}$ and use this together with the fact that e divides $\Phi_1(1)$ to deduce that $\Phi_1(1) = e$ or $(h-1)e$. The latter is impossible since $((h-1)e)^2 > |G|$.]

5

The characterisation of characters

Given a complex-valued class function on a group, we may ask whether or not it is the character of a representation. Certainly a necessary condition is that the values of the function lie in suitable fields of roots of unity, but in this precise form we cannot answer the question. A theorem due to Brauer (1947, 1953) does, however, answer the weaker question as to when a complex-valued class function is a *generalised* character, and a trivial extension gives a criterion for a function to be an irreducible character.

A number of proofs have been given; here we shall follow that due to Brauer and Tate (1955), which is a simplification of Brauer's 1953 proof. A shorter proof has been given by Goldschmidt and Isaacs (1975) but as this depends more on an inductive approach and seems to give less of a feel for the structure of the character ring, we shall give the earlier proof which may be of value if one wishes to examine the relationship between these results and modular character theory. The key steps to the Goldschmidt–Isaacs proof are given as exercises at the end of Section 2.

The approach taken will be to describe the structure of the character ring in terms of Brauer's induction theorem and to obtain Brauer's characterisation of characters as a byproduct. The theorems will be stated in Section 1 and proved in Section 2. Then, in subsequent sections, we will give Brauer's own applications, in which he proved the focal subgroup theorem and determined a splitting field for a finite group. We shall make extensive use of the characterisation of characters in Chapter 6, both in showing that various isometries between spaces of class functions actually carry generalised characters to generalised characters, and also for the explicit construction of certain generalised characters. (See, in particular, Lemma 6.8, Proposition 6.19 and Lemma 6.35 – which the interested reader may wish to regard as exercises! – and also Section 6.6.)

In Section 1 we shall also prove Green's theorem (1955) which states, in effect, that the conditions of Brauer's theorem are necessary.

1 The structure of the character ring

Let p be a prime. Then a group G is said to be a *p-elementary* group if $G = A \times B$ where A is cyclic p'-group and B is a p-group. Let

$$\mathscr{E}_p = \{H \subseteq G \mid H \text{ is } p\text{-elementary}\}$$

and define the set of *elementary subgroups* of G to be the union

$$\mathscr{E} = \bigcup_p \mathscr{E}_p,$$

taken over all prime divisors p of $|G|$.

For H an arbitrary subgroup of G, induction of characters gives rise to a $\mathbb{Z}$-homomorphism

$$\text{ind}: \mathscr{C}\mathscr{h}(H) \to \mathscr{C}\mathscr{h}(G).$$

Let $\mathscr{C}\mathscr{h}^1(H)$ be the subring of $\mathscr{C}\mathscr{h}(H)$ generated by the linear characters. Then we may state Brauer's induction theorem in the following form.

Theorem 1. *The $\mathbb{Z}$-homomorphism*

$$\text{ind}: \bigoplus_{E \in \mathscr{E}} \mathscr{C}\mathscr{h}^1(E) \to \mathscr{C}\mathscr{h}(G)$$

defined by

$$\sum \psi \to \sum \psi^G$$

is surjective.

Put more descriptively, this result asserts that every generalised character (and, in particular, every irreducible character) of G is a $\mathbb{Z}$-linear combination of characters induced from linear characters of elementary subgroups. An elementary group is nilpotent and hence, by Theorem 2.40, monomial. Thus any irreducible character of an elementary group is induced from a linear character of one of its subgroups, which again is elementary, and we could equivalently have taken $\mathscr{C}\mathscr{h}(E)$ in place of $\mathscr{C}\mathscr{h}^1(E)$ for each $E \in \mathscr{E}$; we shall actually prove the theorem in this form.

Theorem 1, or more precisely its proof, realises the following characterisation of characters.

Theorem 2. *Let $\theta \in \mathscr{C}(G; \mathbb{C})$. Then $\theta \in \mathscr{C}\mathscr{h}(G)$ if and only if $\theta|_E \in \mathscr{C}\mathscr{h}(E)$ for every elementary subgroup E of G. If, in addition, $\|\theta\| = 1$ and $\theta(1) > 0$, then θ is an irreducible character.*

Note that the conditions of the final part of this theorem trivially imply that such a generalised character θ is an irreducible character.

Let R be any subring of $\mathbb{C}$ containing the integers. In the final section of Chapter 6, we shall need the generalisation of Theorem 2 to $\mathscr{Ch}_R(G)$, the ring of R-linear combinations of characters of G (as we defined it in Chapter 3). This will be proved at the same time as Theorem 2 but, as the proof will demonstrate, it is actually a simpler statement if R contains sufficiently many roots of unity. The corresponding analogue of Theorem 1 is an easy corollary of Theorem 1 but we shall have no need for it. (See Exercise 1.)

Theorem 2′. *Let $\theta \in \mathscr{C}(G;\mathbb{C})$. Then $\theta \in \mathscr{Ch}_R(G)$ if and only if $\theta|_E \in \mathscr{Ch}_R(E)$ for every elementary subgroup E of G.*

Before embarking on the proofs of these theorems, we shall show the necessity of the hypotheses. It is clearly sufficient to take just one representative of each conjugacy class of elementary subgroups; we shall now prove Green's theorem which asserts that no smaller family will suffice. Specifically, we shall establish the following result.

Theorem 3. *Let G be a group and let $\mathscr{F}$ be a family of subgroups of G such that the $\mathbb{Z}$-homomorphism*

$$\text{ind}: \bigoplus_{F \in \mathscr{F}} \mathscr{Ch}(F) \to \mathscr{Ch}(G)$$

is surjective. Then every elementary subgroup of G has a conjugate which is contained in a member of $\mathscr{F}$.

Proof. Let $E = \langle x \rangle \times P$ be a p-elementary subgroup of G which does not lie in any conjugate of any member of $\mathscr{F}$. Then, without loss, we may suppose that P is a Sylow p-subgroup of $C_G(x)$. Let $F \in \mathscr{F}$ and let $\psi \in \mathscr{Ch}(F)$. Let $S = \mathbb{Z}[\varepsilon]$ where G has exponent m and ε is a primitive mth root of unity. It will suffice to show that $\psi^G(x) \in pS$, for then $1_G(x) \in pS$, which is nonsense.

By taking conjugates if necessary, we may assume that $x \in F$, for otherwise $\psi^G(x) = 0$ and we are done. Now $C_F(x)$ cannot contain a conjugate of P and hence p divides $[C_G(x):C_F(x)]$; this is true also for every conjugate of x that lies in F. Now, by Corollary 2.31,

$$\psi^G(x) = \sum_g \psi(gxg^{-1})$$

where the summation is carried over a set of coset representatives for F in G for which $gxg^{-1} \in F$. The number of times that an H-conjugate of x occurs in this sum is equal to the number of cosets which contain an element of $C_G(x)$, namely $[C_G(x):C_F(x)]$, and this is the case for each element of $x^G \cap H$. Since $\psi(x_0)\in S$ for each conjugate x_0 of x which lies in H, it follows that $\psi^G(x)\in pS$.

Exercises

1. Let χ be an irreducible character of a group G. Show that G has elementary subgroups $E_1,\ldots,E_n$ such that $\chi=\sum_i a_i\varphi_i^G$, where φ_i is an irreducible subgroup of E_i and $a_i\in\mathbb{Z}$ for each i. Deduce that, if R is a subring of $\mathbb{C}$ containing $\mathbb{Z}$, then the map
$$\mathrm{ind}_R\colon \bigoplus_{E\in\mathscr{E}} \mathscr{C}\mathscr{h}_R(E)\to\mathscr{C}\mathscr{h}_R(G)$$
defined analogously to the function ind of Theorem 1 is surjective.
2. Show, by using Theorem 2, that the class functions determined as the irreducible characters of A_5 and PSL(2, 7) (in Sections 2.1 and 2.7, respectively) *are* irreducible characters. Determine the characters as integral linear combinations of the characters induced from elementary subgroups and hence verify the statement of Theorem 3 in these cases.
3. Repeat Exercise 2 for the groups $\mathrm{SL}(2,2^n)$.
4. Let G be a group and let the prime divisors of $|G|$ be divided into two disjoint sets σ and τ. Show that the class function which takes the value $|G|_\sigma$ on elements whose order is not divisible by any prime in σ and 0 otherwise is a generalised character.
[Note that any elementary subgroup of G can be written as a direct product of a σ-group and a τ-group. We shall employ this argument on several occasions in the next chapter.]

2 The proofs of the theorems

Let G be a finite group and let R be any subring of $\mathbb{C}$ containing $\mathbb{Z}$. We shall define the following two subrings of $\mathscr{C}(G;\mathbb{C})$:
$$V_R(G)=\{\theta \mid \theta=\sum r_i\psi_i^G \text{ for } \psi_i\in\mathscr{C}\mathscr{h}(E_i), E_i\in\mathscr{E} \text{ and } r_i\in R\},$$
and
$$U_R(G)=\{\mu \bigm| \mu|_E\in\mathscr{C}\mathscr{h}_R(E) \text{ for each } E\in\mathscr{E}\}.$$
As in the discussion after the statement of Theorem 1, we note that, in the definition of $V_R(G)$, we could have required that the generalised characters ψ_i be linear characters, and equally we could have required that $\psi_i\in\mathscr{C}\mathscr{h}_R(E_i)$. Thus Theorem 1 is merely the assertion that $\mathscr{C}\mathscr{h}_Z(G)=V_Z(G)$, Theorem 2 that $U_Z(G)=\mathscr{C}\mathscr{h}_Z(G)$, and Theorem 2′ that $U_R(G)=\mathscr{C}\mathscr{h}_R(G)$. Clearly $V_R(G)\subseteq\mathscr{C}\mathscr{h}_R(G)\subseteq U_R(G)$ so that we will be able to establish all three theorems by proving that $V_R(G)=U_R(G)$. In order to achieve this, we first note the following.

Lemma 4. *$V_R(G)$ is an ideal of $U_R(G)$.*

Proof. Suppose that $\theta = \sum_i r_i \psi_i^G \in V_R(G)$ with $\psi_i \in \mathscr{C}\hbar(E_i)$. If $\mu \in U_R(G)$, then direct computation (cf. Exercise 19 of Section 2.3) will show that

$$\mu\theta = \sum_i r_i(\mu\psi_i^G) = \sum_i r_i(\psi_i \cdot \mu|_{E_i})^G.$$

Since $\psi_i \cdot \mu|_{E_i} \in \mathscr{C}\hbar_R(E_i)$, it follows that $\mu\theta \in V_R(G)$.

In view of this lemma, it will suffice to show that $1_G \in V_{\mathbb{Z}}(G)$, and we shall do this by 'building' the function 1_G via its constituents on the conjugacy classes of G, using congruences modulo prime divisors of $|G|$. We shall need the following easy group theoretic result.

Lemma 5. *Let $g \in G$ and let p be a prime. Then there are uniquely determined elements g_1 and g_2 in G such that $g = g_1 g_2$ and*

(i) *g_1 is a p-element,*
(ii) *g_2 is a p'-element, and*
(iii) $[g_1, g_2] = 1$.

An element $x \in G$ commutes with g if and only if it commutes with both g_1 and g_2.

Proof. Suppose that g has order $p^\alpha n$ where n is coprime to p. Then there exist integers a, b such that $ap^\alpha + bn = 1$ and, putting $g_1 = g^{bn}$ and $g_2 = g^{ap^\alpha}$, it is easily verified that (i), (ii) and (iii) hold.

Conversely, given such g_1 and g_2, we see that g_1 has order p^α and g_2 has order n. Then $g_1^n = g^n$ and, since $bn \equiv 1 \pmod{p^\alpha}$, $g_1 = g_1 = g^{bn}$, and similarly for g_2.

The final statement is a consequence of the fact that g_1 and g_2 are powers of g.

Definition. Let p be a prime. An element of order coprime to p is said to be *p-regular*. In the situation of Lemma 5, g_1 is called the *p-part* of g and g_2 the *p-regular part* (or *p'-part*).

We can now continue with the proofs of the theorems. Throughout this section, p will denote a prime divisor of $|G|$. As in the proof of Theorem 3, if G has exponent m and ε is a primitive mth root of unity, we shall fix S as the ring $\mathbb{Z}[\varepsilon]$, noting that all character values will lie in S. The next lemma will be applied where a is a p-regular element of G and B a p-subgroup of $C_G(a)$, possibly trivial.

Lemma 6. *Let $E=\langle a\rangle \times B\in\mathscr{E}$. Then there exists $\xi\in\mathscr{C}\ell_S(E)$ such that $\xi(ab)=|\langle a\rangle|$ and $\xi(a^ib)=0$ if $a^i\neq a$, for all $b\in B$.*

Proof. Without loss, we may suppose that $B=1$. If $|\langle a\rangle|=n$ and ω is a primitive nth root of unity, the irreducible characters of E are given by

$$\omega_i(a)=\omega^i, \quad i=1,\dots,n.$$

Then, if ξ is the class function on $\langle a\rangle$ such that $\xi(a)=|\langle a\rangle|$ and $\xi(a^i)=0$ if $a^i\neq a$, we may write $\xi=\sum_i c_i\omega_i$ and compute that

$$c_i=(\xi,\omega_i)=-\omega^i\in S.$$

Lemma 7. *Let a be a p-regular element of G. Then there exists an integer-valued function $\theta\in V_S(G)$ such that $\theta(g)\equiv 1 \pmod p$ if the p-regular part of g is conjugate to a, and $\theta(g)=0$ otherwise.*

Proof. Let B be a Sylow p-subgroup of $C_G(a)$ and put $E=\langle a\rangle\times B$; let ξ be the function defined in Lemma 6. Then ξ and hence also ξ^G are integer-valued. Furthermore, $\xi^G(g)=0$ unless g is conjugate to an element ab for some $b\in B$, in which case the p-regular part of g is conjugate to a. In this case, we have

$$\xi^G(ab)=\frac{1}{|B|}|\{x\in G\,|\,x^{-1}(ab)x=ab' \text{ for some } b'\in B\}|.$$

Now, by Lemma 5,

$$\begin{aligned}\{x\in G\,|\,x^{-1}(ab)x=ab' \text{ for some } b'\in B\}&=\{x\in C(a)\,|\,x^{-1}bx\in B\}\\&=\{x\in C(a)\,|\,b\in xBx^{-1}\}.\end{aligned}$$

Hence

$$|\{x\in G\,|\,x^{-1}(ab)x=ab' \text{ for some } b'\in B\}|=n|N_{C(a)}(B)|$$

where n is the number of Sylow p-subgroups of $C(a)$ which contain b. If $b\neq 1$, then b permutes the Sylow p-subgroups of $C(a)$ which do *not* contain b in orbits of length a power of p. So $n\equiv 1 \pmod p$, and this is true also if $b=1$. Thus

$$\begin{aligned}\xi^G(ab)&=n[N_{C(a)}(B):B]\\&\equiv n\cdot[C(a):N_{C(a)}(B)]\cdot[N_{C(a)}(B):B]\\&\equiv[C(a):B] \pmod p\end{aligned}$$

and hence p does not divide $\xi^G(ab)$. Now $\alpha\xi^G(ab)\equiv 1 \pmod p$ for some integer α and we may take $\theta=\alpha\xi^G$.

Corollary 8. *There exists an integer-valued function $\psi\in V_S(G)$ such that $\psi(g)\equiv 1 \pmod p$ for all $g\in G$.*

Proof. Let $\mathscr{C}_1,\ldots,\mathscr{C}_k$ be the conjugacy classes of p-regular elements of G, and choose θ_i to correspond to an element of $\mathscr{C}_i$ for each i as in the lemma. Then we may take $\psi=\sum_i \theta_i$.

Lemma 9. $1_G \in V_S(G)$.

Proof. If $E=\langle a\rangle$ in Lemma 6 and ξ is as constructed there, then $\xi^G(a)=[C(a):\langle a\rangle]$ and ξ^G vanishes on elements not conjugate to a. Hence, if η is an integer-valued class function whose values are divisible by $|G|$, then $\eta\in V_S(G)$.

Take ψ as in Corollary 8. If $|G|=p^\beta n_p$ where p does not divide n_p, then

$$\psi^{p^\beta}(g)\equiv 1 \pmod{p^\beta}$$

for all $g\in G$. (This congruence holds mod $p^{\beta+1}$.) Put $\theta_p=\psi^{p^\beta}-1_G$. Then $n_p\theta_p(g)$ is divisible by $|G|$ for all $g\in G$ and so $n_p\theta_p\in V_S(G)$. Hence $n_p 1_G\in V_S(G)$. But now the integers n_p as p ranges over all prime divisors of $|G|$ have trivial common divisor; hence there are integers c_p such that $\sum_p c_p n_p=1$ and thus $1_G\in V_S(G)$.

This lemma is sufficient to prove Theorem 2′ in the case that $R\supseteq S$ by Lemma 4 (and also proves the corresponding analogue of Theorem 1), but we are close to completing the proofs of the main theorems in general.

Suppose that the mth cyclotomic polynomial $\varphi_m(x)$ has degree n. Then S has basis $\varepsilon^0=1,\varepsilon,\ldots,\varepsilon^{n-1}$ as a free $\mathbb{Z}$-module. Let H be any subgroup of G; then we claim that $1,\varepsilon,\ldots,\varepsilon^{n-1}$ are linearly independent over $\mathscr{Ch}_{\mathbb{Z}}(H)$ in $\mathscr{Ch}_S(H)$. For if H has irreducible characters $\varphi_1,\ldots,\varphi_q$ and

$$\sum_{i=0}^{n-1}\varepsilon^i\left(\sum_{j=1}^{q}c_{ij}\varphi_j\right)=0$$

for integers c_{ij}, then

$$\sum_{j=1}^{q}\left(\sum_{i=0}^{n-1}\varepsilon^i c_{ij}\right)\varphi_j=0.$$

Now $\sum_i c_{ij}\varepsilon^i=0$ for each j, whence all the coefficients $c_{ij}=0$.

By Lemma 9, there are scalars $c_i\in S$ and generalised characters ψ_i of elementary subgroups E_i such that

$$1_G=\sum_i c_i\psi_i^G.$$

Then

$$1_G=\sum_i \tilde{\psi}_i^G$$

where $\tilde{\psi}_i \in \mathscr{C}\ell_S(E_i)$. However, for each i, we may write

$$\tilde{\psi}_i = \sum_j \varepsilon^j \varphi_{ij}$$

where $\varphi_{ij} \in \mathscr{C}\ell_Z(E_i)$, and then

$$1_G = \sum_{i,j} \varepsilon^j \varphi_{ij}^G.$$

Since $\varphi_{ij}^G \in \mathscr{C}\ell_Z(G)$ for all i,j and $1, \varepsilon, \ldots, \varepsilon^{n-1}$ are linearly independent over $\mathscr{C}\ell_Z(G)$, we may deduce that

$$1_G = \sum_i \varphi_{i0}^G \in V_Z(g).$$

Exercises

(The goal of Exercises 1–8 will be to develop the Goldschmidt–Isaacs proof of Theorems 1 and 2.)

1. A group H is said to be (p-)*hyperelementary* if H has a cyclic normal p-complement. Show that subgroups of p-hyperelementary subgroups are p-hyperelementary.
2. Let G be any group and let $x \in G$. Let C be the normal p-complement of $\langle x \rangle$ (possibly trivial) and let H be a subgroup of $N(C)$ such that H/C is a Sylow p-subgroup of $N(C)/C$. Show that H is p-hyperelementary and that $x \in H$. Show that in the permutation action of G on the right cosets of H in G, a coset of H fixed by x lies in N. Deduce that p does not divide $(1_H)^G(x)$.
3. Show that the product of two permutation characters of a group G is a permutation character. Deduce that the set of integral combinations of permutation characters of G form a subring $\mathscr{P}\mathscr{C}\ell(G)$ of $\mathscr{C}\ell(G)$.
4. Let $\mathscr{H}$ be the set of hyperelementary subgroups of a group G. Let $\mathscr{P}\mathscr{C}\ell(G, \mathscr{H})$ be the subring of $\mathscr{P}\mathscr{C}\ell(G)$ generated by the permutation characters $(1_H)^G$ for $H \in \mathscr{H}$. Show that $\mathscr{P}\mathscr{C}\ell(G, \mathscr{H})$ is an ideal of $\mathscr{P}\mathscr{C}\ell(G)$.
5. (Banaschewski) Let $\mathscr{R}$ be a ring of integer-valued functions on a nonempty finite set T, with addition and multiplication defined pointwise. If the constant function $1_T \notin \mathscr{R}$, show that there exists a prime p and $t \in T$ such that p divides $f(t)$ for all $f \in \mathscr{R}$.
   [Hint. Suppose that $I_t = \{f(t) \mid f \in \mathscr{R}\} = \mathbb{Z}$ for all $t \in T$. Pick $f_t \in \mathscr{R}$ with $f_t(t) = 1$ and consider $\prod_t (f_t - 1_T)$.]
6. (L. Solomon) Use the preceding exercises to show that $1_G \in \mathscr{P}\mathscr{C}\ell(G, \mathscr{H})$.
7. Let G be a p-hyperelementary group with normal p-complement N.

Let λ be an irreducible character of N such that $\ker \lambda$ contains $Z(G)$ and $I(\lambda) = G$. Show that $\lambda = 1_N$.

8. Prove that $1_G \in V_Z(G)$ by induction on $|G|$.
[Hint. Exercise 6 reduces the problem to the case that G is hyperelementary but not elementary. If $G = PN$ with N a normal p-complement, let $E = P \times C_N(P)$. Then $E \neq G$. Use Mackey's decomposition theorem to examine $(1_H)^G|_N$ and use Exercise 7 to show that each nonprincipal constituent of $(1_H)^G$ is induced from a proper subgroup of G.]
9. Let G be a group and H a Hall π-subgroup of G. Suppose that

(a) whenever two elements of H are conjugate in G, they are already conjugate in H, and
(b) if $x \in H$ and $x \notin Z(G)$, then $C_H(x)$ is a Hall π-subgroup of $C_G(x)$ and $C_G(x)$ has a normal subgroup $I(x)$ such that $C_G(x) = C_H(x)I(x)$ and $C_H(x) \cap I(x) = 1$.

Establish the following.

(i) Any π-element of G (element whose order is divisible only by primes in π) is conjugate to an element of H.
(ii) If $x \in G$, then x can be written uniquely as a product of a π-element (the *π-part* of x) and a π'-element which commute.
(iii) Let θ be the inflation to H of an irreducible character of $H/(H \cap Z(G))$ of degree d. For $x \in G$, define

$$\varphi(x) = \theta(y) - d$$

where y is an element of h conjugate to the π-part of x. Show that φ is well defined and use Theorem 2 to show that $\varphi \in \mathscr{C}\mathscr{h}(G)$.
(iv) Show that there is an irreducible character χ of G such that $\varphi = \chi - d \cdot 1_G$.
(v) Deduce, by considering the intersection of the kernels of all characters χ obtained in this way, that G possesses a normal subgroup N such that $G = HN$ and $H \cap N = 1$.

[Part (ii) is a direct generalisation of Lemma 5. To show parts (iv) and (v), generalise the argument of the proof of Frobenius' theorem; the subgroup N is called a *normal π-complement*. This theorem is due to Suzuki (1963).]

3 The focal subgroup theorem

The existence of a proper abelian homomorphic image of a group is equivalent to the existence of a nonprincipal linear character. We shall therefore begin with a characterisation of such characters.

Proposition 10. *Let $\theta \in \mathscr{C}(G, \mathbb{C})$. Then θ is a linear character of G if and only if the following conditions hold:*

(i) *θ is not identically zero, and*
(ii) *$\theta(gh) = \theta(g)\theta(h)$ whenever either*
  (a) *g and h are commuting elements of coprime order, or*
  (b) *$\langle g, h \rangle$ has prime power order.*

Proof. The necessity of these conditions is obvious. Conversely, since $\theta \not\equiv 0$, it follows from (ii) that $\theta(1) = 1$ and then that, whenever P is a p-subgroup, $\theta|_P$ is a homomorphism from P into $\mathbb{C}^\times$ and hence a linear character. By (ii)(a), $\theta|_N$ is now a linear character for any nilpotent subgroup N and, in particular, $\theta|_E$ is a linear character for any elementary subgroup. Thus $\theta \in \mathscr{C}\mathscr{h}(G)$; also $|\theta(g)| = 1$ for all $g \in G$ and hence $\|\theta\| = 1$. So θ is a linear character of G.

Theorem 11 (Focal subgroup theorem). *Let P be a Sylow p-subgroup of the finite group G and let*

$$P_0 = \langle xy^{-1} | x, y \in P,\ x, y \text{ conjugate in } G \rangle.$$

Then $P \cap G' = P_0$.

Proof. As in the earlier proof (Theorem 1.46), we have

$$P' \subseteq P_0 \subseteq P \cap G'$$

and, in particular, that P/P_0 is abelian.

Let $p_1, \ldots, p_m$ be the distinct prime divisors of $|G|$ and, for each i, let P_i be a fixed Sylow p_i-subgroup; we may suppose that $P = P_1$. If $g \in G$, then $g = g_1 \cdots g_m$ where g_i is a p_i-element and $g_1, \ldots, g_m$ commute pairwise, and g_1 is conjugate to some element $h_1 \in P_1$. If g_1 is also conjugate to some element $\tilde{h}_1 \in P_1$, then $h_1 \tilde{h}_1^{-1} \in P_0$. Thus, if λ is a linear character of P with $P_0 \subseteq \ker \lambda$, then $\lambda(h_1) = \lambda(\tilde{h}_1)$ and the function θ given by

$$\theta(g) = \lambda(h_1)$$

is well defined.

Clearly θ is a class function and the hypotheses of Proposition 10 are satisfied; so θ is a linear character of G. If $x \in P - P_0$, there is a linear character λ of P such that $\lambda(x) \neq 1$; then $\theta(x) \neq 1$ and $x \notin P \cap G'$. Hence $P \cap G' \subseteq P_0$.

4 A splitting field

Let G be a finite group. In Theorem 1.14, we showed that there is an algebraic number field which is a splitting field for G. If K is such a field, then it is clearly necessary that K contain all the values of the irreducible characters of G, but from our discussion of the Frobenius–Schur indicator in Chapter 2, we know that this is certainly not sufficient. The general question of the relationship between splitting fields and the fields of character values and, more specifically, the relationship between the field of a single irreducible character and the smallest fields in which it may be realised (namely, the Schur index) are ones which we shall not consider in this book (but see, for example, Curtis and Reiner [CR II; §74]); what we shall restrict ourselves to here is the explicit example of a splitting field given by Brauer as an application of Theorem 1, and it is evident (for example, by considering abelian groups) that no better general result can exist.

Theorem 12. *Let G be a finite group of exponent m and let ε be a primitive mth root of unity. Then $\mathbb{Q}(\varepsilon)$ is a splitting field for G.*

Proof. Let $F = \mathbb{Q}(\varepsilon)$. Then a linear character φ of an elementary subgroup of G is realisable over F as then is the induced character φ^G. (See Corollary 1.30, though this can be seen directly in this case.)

Now let χ be an irreducible character of G. By Theorem 1,

$$\chi = \sum a_i \varphi_i^G$$

where the φ_i are linear characters of elementary subgroups of G. Then

$$\chi = \psi_1 - \psi_2$$

where ψ_1 and ψ_2 are characters of G afforded by FG-modules L_1 and L_2 respectively. If M is a $\mathbb{C}G$-module affording χ, then

$$L_1^{\mathbb{C}} \cong_{\mathbb{C}G} L_2^{\mathbb{C}} \oplus M.$$

It will therefore suffice to exhibit an FG-module N such that $M \cong_{\mathbb{C}G} N^{\mathbb{C}}$. We shall do this by means of a general lemma.

Lemma 13. *Let G be a group and F a subfield of $\mathbb{C}$. Let L_1 and L_2 be FG-modules and M a $\mathbb{C}G$-module such that*

$$L_1^{\mathbb{C}} \cong_{\mathbb{C}G} L_2^{\mathbb{C}} \oplus M.$$

Then there exists an FG-module N such that $M \cong_{\mathbb{C}G} N^{\mathbb{C}}$.

Proof. If $L_2 = 0$, there is nothing to prove, so assume that modules are chosen so that the conclusion is false, with L_2 of smallest possible F-dimension. Let L_0 be an irreducible submodule of L_2. Then $\mathrm{Hom}_{FG}(L_0, L_2) \neq 0$ and so, by Theorem 1.12,

$$\mathrm{Hom}_{\mathbb{C}G}(L_0^{\mathbb{C}}, L_2^{\mathbb{C}}) \cong \mathbb{C} \otimes_F \mathrm{Hom}_{FG}(L_0, L_2) \neq 0.$$

Thus $\mathrm{Hom}_{\mathbb{C}G}(L_0^{\mathbb{C}}, L_1^{\mathbb{C}}) \neq 0$ and hence, by Theorem 1.12 again,

$$\mathrm{Hom}_{FG}(L_0, L_1) \neq 0.$$

Since L_0 is irreducible and L_1 is completely reducible, there is an FG-module P such that $L_1 \cong L_0 \oplus P$. If $L_2 = L_0 \oplus L_0'$, then $P^{\mathbb{C}} \cong (L_0')^{\mathbb{C}} \oplus M$; now, $M = N^{\mathbb{C}}$ for some FG-module N by the minimality of $\dim L_2$.

6

Isometries

In this final chapter, we shall extend the ideas of our earlier discussions. In Chapters 3 and 4, we were essentially studying situations where T.I.-sets exist. In practice, this is often too restrictive a condition in that group theoretic reductions are unable to achieve so tight a configuration. This is most notable in the Odd Order Theorem where special isometries had to be constructed in order to study groups containing subgroups with so-called tamely imbedded subsets. The arguments there were generalised by Dade (1964), and in the first section of this chapter we shall give an account of the isometry that he constructed.

His particular hypotheses, whilst sufficient for the proof of the Odd Order Theorem, are restrictive, and we shall consider various ways in which they may be relaxed. In a special situation, Brauer's characterisation of characters will prove the desired result: it is often not difficult to construct isometries of *class functions*, but then far more complicated to establish that the constructed map actually sends a generalised character into a generalised character. Further refinements depend on the use of block theory. The necessary facts can be stated in terms of ordinary characters alone, rather than in terms of the modular representation theory on which they depend, and this we shall do in Section 2, omitting those proofs which depend on modular representation theory, contrary to the intention of this book (in all other respects) to be self-contained – it is our intention at this point to give the flavour of what may be achieved, rather than to give an exhaustive account. This will permit us to examine groups with certain Sylow 2-subgroups; specifically, we shall be able to show that groups having homocyclic Sylow 2-subgroups of rank 2 and order greater than 4 cannot be simple and to give a proof of the Brauer–Suzuki theorem for groups with a quaternion Sylow 2-subgroup of order 8.

In Section 4, we shall establish some refinements of the method of exceptional characters, showing the way in which it respect blocks. Finally, we will discuss isometries constructed by Reynolds (1968) and Robinson (1985) in Section 7.

1 Dade's isometry

Let G be a group and let π be a set of primes. An element $x \in G$ is called a *π-element* if every prime divisor of the order of x lies in π. The complementary set of primes is denoted by π'.

Lemma 1. *Let $g \in G$. Then there are uniquely determined elements g_π and $g_{\pi'}$ in G such that $g = g_\pi g_{\pi'}$ and*

(i) *g_π is a π-element,*
(ii) *$g_{\pi'}$ is a π'-element, and*
(iii) $[g_\pi, g_{\pi'}] = 1$.

Proof. Suppose that g has order mn where m is a π-number and n is a π'-number. Then there exist integers a, b such that $am + bn = 1$ and the proof follows that of Lemma 5.5 (which is the special case $\pi = \{p\}$).

In this situation, the elements g_π and $g_{\pi'}$ are called the *π-part* and *π'-part* of g, respectively. If x is a π-element, then the *π-section* $\mathscr{S}_\pi^G(x)$ of x is the set

$$\mathscr{S}_\pi^G(x) = \{g \in G \mid g_\pi \sim x\}.$$

We shall suppress the superscript if no ambiguity can occur. If $\pi = \{p\}$, we refer simply to *p-sections*, and write $\mathscr{S}_p(x)$. The following properties of π-sections are easily verified by using Lemma 1.

Lemma 2. *$\mathscr{S}_\pi(x)$ is a union of conjugacy classes of G. Furthermore, if $g, h \in \mathscr{S}_\pi(x)$ and $g_\pi = h_\pi$, then $g \sim h$ if and only if $g_{\pi'} \sim_{C(g_\pi)} h_{\pi'}$ and*

$$|\{g \in \mathscr{S}_\pi(x) \mid g_\pi = x\}| = \text{number of } \pi'\text{-elements in } C_G(x).$$

We may now discuss the hypothesis studied by Dade and his construction of an isometry.

Hypothesis I. G is a finite group, N is a subgroup of G and A is a union of conjugacy classes of π-elements of N. The following conditions hold.

(i) If $a_1, a_2 \in A$ are conjugate in G, then they are conjugate in N.
(ii) If $a \in A$, then $C_G(a) = C_N(a) \cdot I(a)$ where $I(a)$ is a normal Hall π'-subgroup of $C_G(a)$, and $C_N(a) \cap I(a) = 1$.

We observe that condition (i) tells us that N controls the G-fusion of elements in A while if, in addition, $I(a) = 1$ for all $a \in A$, the set A consists of a set of special classes as defined in Chapter 3. Indeed, if N is a Frobenius group with an abelian Frobenius kernel H and $N = N_G(H)$, then an easy application of Burnside's transfer theorem shows that Hypothesis I is satisfied with $A = H^\#$.

However, we shall want eventually to weaken such conditions. Notice that it follows immediately from condition (ii) that $C_N(a)$ is a Hall π-subgroup of $C_G(a)$. Thus, for example, if H were an abelian Hall π-subgroup of G with $N = H\langle y \rangle = N_G(H)$ where y is an element of order 4 inverting every element of H, then we could take $A = H^\#$ to satisfy condition (i). Again, Burnside's transfer theorem will apply and show that $C_G(a) = C_N(a) \cdot I(a)$ for all $a \in A$, where $I(a)$ is a normal Hall π'-subgroup of $C_G(a)$, but in this case $C_N(a) \cap I(a) = \langle y^2 \rangle$. So Hypothesis I is not satisfied, although we are clearly sufficiently close to (and even might be in) a T.I.-set situation that we might hope for corresponding results. We shall return to this later in this section.

Let N be a subgroup of G and let A be a union of conjugacy classes of π-elements of N. We extended the notation of Chapter 3 by letting $\mathscr{C}_N(A)$ denote the subspace of $\mathscr{C}(N; \mathbb{C})$ consisting of those class functions on N vanishing outside A; as before, $\mathscr{M}_N(A)$ will denote the submodule of generalised characters of N vanishing outside A.

Now assume that Hypothesis I holds. In view of condition (i), we may define a map $\tau: \mathscr{C}_N(A) \to \mathscr{C}(G; \mathbb{C})$ by

$$\theta^\tau(g) = \begin{cases} \theta(a) & \text{if } g \in \mathscr{S}_\pi(a) \text{ for } a \in A, \\ 0 & \text{otherwise,} \end{cases} \tag{1.1}$$

for each $\theta \in \mathscr{C}_N(A)$. In the case that π consists of all primes, then this map coincides with the usual induction of class functions.

Theorem 3. *Assume that Hypothesis I holds and that τ is given by (1.1). Then τ is an isometry. Furthermore,*

(i) *if $\alpha \in \mathscr{C}(G; \mathbb{C})$ and α is constant on the π-section $\mathscr{S}_\pi(a)$ for each $a \in A$, then $(\theta^\tau, \alpha)_G = (\theta, \alpha|_N)_N$, and*

(ii) *if θ is a generalised character, then θ^τ is a generalised character of G.*

Remark. Statement (i) can be viewed as a partial analogue of the Frobenius reciprocity formula; in particular, it holds for $\alpha = 1_G$. It forms the basis for an isometry due to Reynolds (1967) given by π-induction. (See Exercises 7 and 8.)

Proof. Let $\mathscr{C}_1, \ldots, \mathscr{C}_m$ be the conjugacy classes of N that lie in A, and let $x_1, \ldots, x_m$ be representatives of those classes. If $\theta \in \mathscr{C}_N(A)$, then θ^τ vanishes except on the π-sections $\mathcal{S}_\pi(x_1),\ldots,\mathcal{S}_\pi(x_m)$ and, if $\varphi\in\mathcal{C}_N(A)$ also, we may simply compute $(\theta^\tau,\varphi^\tau)_G$ to see that

$$(\theta^\tau,\varphi^\tau)_G=\frac{1}{|G|}\sum_{i=1}^m\left(\sum_{g\in\mathcal{S}_\pi(x_i)}\theta^\tau(g)\overline{\varphi^\tau(g)}\right)$$
$$=\frac{1}{|G|}\sum_{i=1}^m\frac{|G|}{|C_G(x_i)|}\cdot|I(x_i)|\theta(x_i)\overline{\varphi(x_i)} \qquad (1.2)$$

by Lemma 2. But then

$$(\theta^\tau,\varphi^\tau)_G=\sum_{i=1}^m\frac{1}{|C_N(x_i)|}\theta(x_i)\overline{\varphi(x_i)}=\frac{1}{|N|}\sum_{x\in N}\theta(x)\overline{\varphi(x)}=(\theta,\varphi)_N.$$

So τ is an isometry.

Now, for any α as in (i), define $\alpha_0\in\mathcal{C}_N(A)$ by

$$\alpha_0(g)=\begin{cases}\alpha(g) & \text{if } g\in A,\\ 0 & \text{otherwise.}\end{cases}$$

Then α_0^τ agrees with α on the π-sections $\mathcal{S}_\pi(x_1),\ldots,\mathcal{S}_\pi(x_m)$ and hence

$$(\theta^\tau,\alpha)_G=(\theta^\tau,\alpha_0^\tau)_G=(\theta,\alpha_0)_N=(\theta,\alpha|_N)_N,$$

the last equality holding since $\alpha|_A=\alpha_0|_A$.

Suppose, now, that $\theta\in\mathcal{M}_N(A)$. In order to show that $\theta^\tau\in\mathcal{Ch}(G)$, it is enough to show that $\theta^\tau|_E\in\mathcal{Ch}(E)$ for every elementary subgroup E of G. But if $E=X\times Y$ where X is a π-group and Y is a π'-group, $\theta^\tau(xy)=\theta^\tau(x)$ for $x\in X$ and $y\in Y$ since θ^τ is constant on π-sections, so that it is enough to show that $\theta^\tau|_X\in\mathcal{Ch}(X)$. Thus it will suffice to exhibit a generalised character ξ of G such that θ^τ and ξ agree on every π-element of G.

The function that we shall consider is

$$\xi=-\sum_B\frac{(-1)^{|B|}}{[N:N_N(B)]}(\theta_{(B)})^G \qquad (1.3)$$

where the summation runs over all nonempty subsets B of A, and $\theta_{(B)}$ is a generalised character of a certain subgroup to be associated with B, having the property that $(\theta_{(B)})^G=(\theta_{(B^n)})^G$ for any $n\in N$. Since B has precisely $[N:N_N(B)]$ distinct conjugates in N, all of them subsets of A, it follows that $\xi\in\mathcal{Ch}(G)$.

In order to define $\theta_{(B)}$ and complete the proof of Theorem 3, we shall need the following consequence of Hypothesis I.

Lemma 4. *Let B be a nonempty subset of A, and let*

$$I(B)=\bigcap_{b\in B}I(b).$$

Then $I(B)$ is a Hall π'-subgroup of $C_G(B)$. In particular, $N_N(B)$ normalises $I(B)$ and $N_N(B)\cap I(B)=1$.

Proof. For each $b\in B$, the subgroup $I(B)$ consists of all π'-elements in $C_G(b)$, and hence $I(B)$ consists of all π'-elements in $C_G(B)$. Thus $I(B)$ is a Hall π'-subgroup of $C_G(B)$ and the remainder of the lemma is clear since $N_N(B)\cap I(B)\subseteq C_N(B)\cap I(B)=1$.

We now define $\theta_{(B)}$ as the inflation to $N_N(B)\cdot I(B)$ of $\theta|_{N_N(B)}$. Since $I(B)$ is a π'-group, $\theta_{(B)}$ vanishes except on elements with π-part in A: hence $(\theta_{(B)})^G$ vanishes outside $\mathscr{S}_\pi(x_1)\cup\cdots\cup\mathscr{S}_\pi(x_m)$. So it will be enough to check the values of θ^τ and ξ on A.

Let $a\in A$. Then

$$(\theta_{(B)})^G(a)=\frac{1}{|N_N(B)\cdot I(B)|}\sum_g \theta_{(B)}(gag^{-1})$$

where the summation is taken over those elements g for which $gag^{-1}\in N_N(B)\cdot I(B)$. If a_1 is a fixed conjugate of a in $N_N(B)$, then the number of conjugates of a lying in the coset $a_1I(B)$ is $[I(B):C_{I(B)}(a_1)]$. Also, for any such element a_1,

$$|\{g\in G|gag^{-1}=a_1\}|=|C_G(a)|,$$

and hence

$$\begin{aligned}(\theta_{(B)})^G(a)&=\sum_{a_1}\frac{|C_G(a)|\cdot[I(B):C_{I(B)}(a_1)]}{|N_N(B)\cdot I(B)|}\theta(a)\\ &=\theta(a)\cdot|C_G(a)|\sum_{a_1}\frac{1}{|N_N(B)|\cdot|C_{I(B)}(a_1)|}\end{aligned}$$

where the summation is taken over all conjugates a_1 of a lying in $N_N(B)$. Returning to (1.3), we may obtain a Möbius-like formula

$$\xi(a)=-\theta(a)\cdot|C_G(a)|\sum_B\sum_{a_1}\frac{(-1)^{|B|}}{|N|\cdot|C_{I(B)}(a_1)|}$$

where we may view the summation over pairs (B,a_1). But then, by reversing the order of summation, we obtain the formula

$$\xi(a)=-\frac{\theta(a)\cdot|C_G(a)|}{|N|}\sum_{a_1}\sum_B\frac{(-1)^{|B|}}{|C_{I(B)}(a_1)|} \tag{1.4}$$

where the first summation is taken for $\{a_1\in A|a_1\sim a\}$.

Now, if $a_1\notin B$,

$$C_{I(B)}(a_1)=I(B)\cap C_G(a_1)=I(B)\cap I(a_1)=I(B\cup\{a_1\})=C_{I(B\cup\{a_1\})}(a_1)$$

since $I(a_1)$ consists precisely of the π'-elements of $C_G(a_1)$. Hence the contributions to the inner sum in (1.4) from B and from $B \cup \{a_1\}$ cancel, leaving only a term from $\{a_1\}$, and so

$$\xi(a) = \frac{\theta(a)\cdot|C_G(a)|}{|N|}\sum_{a_1}\frac{1}{|I(a_1)|}$$
$$= \frac{\theta(a)\cdot|C_N(a)|\cdot|I(a)|}{|N|}\cdot\frac{[N:C_N(a)]}{|I(a)|} = \theta(a) = \theta^\tau(a).$$

This completes the proof of Theorem 3.

A close examination of this proof shows that one obstruction to a head-on attack via the characterisation of characters lies in the possibility that the π-group X may contain elements of $N - A$ which, nevertheless, may be conjugate in G to elements of A. Then the restriction of θ^τ to X does not help us, though this problem does not occur if A consists of all nonidentity π-elements in N. In such a situation, we may proceed to a direct verification that $\theta^\tau|_X \in \mathscr{Ch}(X)$ if $\theta \in \mathscr{M}_N(A)$ as follows.

Take x to be an element of prime order in $Z(X)$. If A consists of all nonidentity π-elements in N, then N must contain a Sylow p-subgroup of G for each $p \in \pi$: so we may suppose that $x \in A$. Now $X \subseteq C_G(x) = C_N(x)\cdot I(x)$. By the Schur–Zassenhaus theorem, since we may replace X by any conjugate, we may suppose that $X \subseteq N$. But then $\theta^\tau|_X = \theta|_X$.

Further examination of this argument in fact shows that, in the case where A consists of all nonidentity π-elements of N, we can achieve both conclusions (i) and (ii) of Theorem 3 without the full recourse to condition (ii) of Hypothesis I: it would be enough for the above argument to be able to identify X with some section of N. However, we cannot show that τ is an isometry without some condition to ensure that we can count elements correctly in (1.2). One way to achieve this, which will cover the problem discussed after the statement of Hypothesis I and which is adequate for many applications, is the following where, for any group H, we denote the largest normal π'-subgroup of H by $O_{\pi'}(H)$.

Hypothesis II. (i) Condition (i) of Hypothesis I holds.

(ii) If $a \in A$, then $C_G(a)$ has a normal π-complement and, in addition, $C_G(a) = C_N(a)\cdot O_{\pi'}(C_G(a))$ and $C_N(a) \cap O_{\pi'}(C_G(a)) \subseteq O_{\pi'}(N)$.

Observe that this does, in fact, generalise Hypothesis I.

Lemma 5. *Assume that G satisfies Hypothesis II. Let* $\tilde{N} = N/O_{\pi'}(N)$ *and put* $\tilde{a} = aO_{\pi'}(N)$, *etc. Then*

$$C_{\tilde{N}}(\tilde{a}) \cong C_N(a)/C_N(a) \cap O_{\pi'}(C_G(a)).$$

Proof. Since $O_{\pi'}(N)\cap C_N(a)\subseteq O_{\pi'}(C_N(a))$ in general, this inclusion becomes equality under Hypothesis II. Now, under the natural homomorphism from N onto $\tilde{N}$, $C_N(a)$ maps into $C_{\tilde{N}}(\tilde{a})$ and, since a has order coprime to $|O_{\pi'}(N)|$, this map is surjective.

As a particular consequence, we note that $C_{\tilde{N}}(\tilde{a})$ is always a π-group. Let $\tilde{A}$ be the image of A in $\tilde{N}$. Then there is a one-to-one correspondence between the N-conjugacy classes in A and the $\tilde{N}$-conjugacy classes in $\tilde{A}$, and we may construct a map σ in place of τ in the following manner.

Let $\mathscr{C}_{\tilde{N}}(\tilde{A})$ and $\mathscr{M}_{\tilde{N}}(\tilde{A})$ have the same meanings in $\mathscr{C}(\tilde{N};\mathbb{C})$ as $\mathscr{C}_N(A)$ and $\mathscr{M}_N(A)$ did in $\mathscr{C}(N;\mathbb{C})$. Assume that Hypothesis II holds. Then we may regard $\mathscr{C}_{\tilde{N}}(\tilde{A})$ as a subspace of $\mathscr{C}(N;\mathbb{C})$ and $\mathscr{M}_{\tilde{N}}(\tilde{A})$ as a submodule of $\mathscr{C}\mathscr{h}(N)$ by inflation. Now define a map $\sigma:\mathscr{C}_{\tilde{N}}(\tilde{A})\to\mathscr{C}(G;\mathbb{C})$ by

$$\theta^\sigma(g)=\left\{\begin{array}{ll}\theta(a) & \text{if } g\in\mathscr{S}_\pi(a) \text{ for } a\in A,\\ 0 & \text{otherwise,}\end{array}\right\} \tag{1.5}$$

for each $\theta\in\mathscr{C}_{\tilde{N}}(\tilde{A})$. We observe that this is well defined since, in view of Lemmas 2 and 5, θ is already constant on the π-sections of elements of A in N.

Theorem 6. *Assume that Hypothesis II holds and that σ is defined as in* (1.5). *Assume further that A consists of all nonidentity π-elements of N. Then the same conclusions hold for σ as did for τ in Theorem 3.*

Proof. We imitate the proof of Theorem 3. Let $\mathscr{C}_1,\dots,\mathscr{C}_m$ be the conjugacy classes of N which lie in A, and let $x_1,\dots,x_m$ be representatives. For $\theta,\varphi\in\mathscr{C}_{\tilde{N}}(\tilde{A})$, we obtain, in place of (1.2),

$$(\theta^\sigma,\varphi^\sigma)_G=\frac{1}{|G|}\sum_{i=1}^m\frac{|G|}{|C_G(x_i)|}\cdot|O_{\pi'}(C_G(x_i))|\cdot\theta(x_i)\overline{\varphi(x_i)}. \tag{1.6}$$

Now, by Lemma 5,

$$[C_G(x_i):O_{\pi'}(C_G(x_i))]=|C_{\tilde{N}}(\tilde{x}_i)|,$$

and so (1.6) can be rewritten as

$$\begin{aligned}(\theta^\sigma,\varphi^\sigma)_G&=\sum_{i=1}^m\frac{1}{|C_{\tilde{N}}(\tilde{x}_i)|}\theta(\tilde{x}_i)\overline{\varphi(\tilde{x}_i)}=\frac{1}{|\tilde{N}|}\sum_{\tilde{x}\in\tilde{N}}\theta(\tilde{x})\overline{\varphi(\tilde{x})}\\&=(\theta,\varphi)_{\tilde{N}}=(\theta,\varphi)_N.\end{aligned}$$

If $\alpha\in\mathscr{C}(G;\mathbb{C})$ and α is constant on each of the π-sections $\mathscr{S}_\pi(x_i)$, then we define $\alpha_0\in\mathscr{C}_{\tilde{N}}(\tilde{A})$ by

$$\alpha_0(\tilde{g})=\begin{cases}\alpha(g) & \text{if } \tilde{g}\in\tilde{A},\\ 0 & \text{otherwise.}\end{cases}$$

As in the proof of Theorem 3, α_0^σ agrees with α on the π-sections $\mathscr{S}_\pi(x_1), \ldots, \mathscr{S}_\pi(x_m)$, and we proceed as before.

Suppose that $\theta \in \mathscr{M}_{\tilde{N}}(\tilde{A})$. Then it will again be sufficient in order to prove that θ^σ is a generalised character to show that $\theta^\sigma|_X \in \mathscr{C}\mathscr{h}(X)$ for any nilpotent π-subgroup X of G. Since A consists of all nonidentity π-elements in N, we know that N must contain a Sylow p-subgroup of G for each $p \in \pi$. For any element x of prime order in $Z(X)$, we may suppose that $x \in A$ and hence that

$$X \subseteq C_G(x) = C_N(x) \cdot O_{\pi'}(C_G(x)).$$

Since X is a π-group, Lemma 5 gives an injection

$$X \to \tilde{X} \subseteq C_{\tilde{N}}(\tilde{x}).$$

Let $\tilde{\theta}$ be the class function on $\tilde{X}$ corresponding to $\theta^\sigma|_X$ under this injection. Then visibly $\tilde{\theta} = \theta|_{\tilde{X}}$ and hence $\tilde{\theta} \in \mathscr{C}\mathscr{h}(\tilde{X})$. But then $\theta^\sigma|_X \in \mathscr{C}\mathscr{h}(X)$ as we require.

We shall complete this section with a normal complement theorem which generalises the argument used to prove Frobenius' theorem and also the theorem of Suzuki (1963) which appeared as Exercise 9 in Section 5.2. (His result is the special case where L is a Hall π-subgroup and $L_0 = L \cap Z(G)$.) Other versions can be proved, corresponding to the different isometries that we shall discuss.

Theorem 7. *Let G be a finite group and let L be a subgroup of G. Suppose that π is a set of primes and that L_0 is a normal subgroup of L such that*

(i) *L/L_0 is a π-group,*
(ii) *any two π-elements of $L - L_0$ which are conjugate in G are conjugate in L, and*
(iii) *if x is a π-element in $L - L_0$, then $C_G(x) = C_L(x) \cdot I(x)$ where $I(x)$ is a normal Hall π'-subgroup of $C_G(x)$ and $C_L(x) \cap I(x) = 1$.*

Then G has a normal subgroup G_0 such that $G = G_0 L$ and $G_0 \cap L = L_0$.

We remark that an easy group theoretical argument shows that the subgroup G_0 is uniquely determined; G_0 is often called a *relative normal complement* or a *normal complement over L_0*.

The particular hypotheses of the theorem imply that every element of $L - L_0$ is a π-element. However, in the proof we shall refer instead to the set of π-elements in $L - L_0$ to give a clear indication of the generalisations which will become possible by using the better isometries which are discussed in Section 7; indeed, one can see immediately that if $L_0 = O_{\pi'}(L)$, then an analogous result will hold under the centraliser condition of Hypothesis II, using the isometry of Theorem 6.

Proof of Theorem 7. Let A be the set of π-elements in $L-L_0$; then Hypothesis I is satisfied with $N=L$. Let τ be the isometry defined in Theorem 3 and let θ be a nonprincipal irreducible character of L of degree d having L_0 in its kernel. Then $(\theta-d\cdot 1_L)^\tau \in \mathscr{C}\ell(G)$ and $\|(\theta-d\cdot 1_L)^\tau\|^2 = 1+d^2$. Also, by Theorem 3,

$$((\theta-d\cdot 1_L)^\tau, 1)_G = -d.$$

So there exists a nonprincipal irreducible character χ of G such that

$$(\theta-d\cdot 1_L)^\tau = \chi - d\cdot 1_G$$

since $(\theta-d\cdot 1_L)^\tau(1)=0$.

Let G_0 be the intersection of the kernels of all the characters χ of G which can be obtained in this way. For $x\in L-L_0$, we have $\chi(x)=\theta(x_\pi)$; hence $G_0\cap L\subseteq L_0$ and so $[G:G_0]\geqslant [L:L_0]$. It will now suffice to prove equality, for then $G_0\cap L=L_0$ and $G=G_0L$.

G_0 consists precisely of those π-sections of G which do not meet $L-L_0$. By Lemma 2,

$$|\mathscr{S}_\pi^G(x)| = [G:C_G(x)]\cdot|I(x)| = [G:C_L(x)]$$

and

$$|\mathscr{S}_\pi^L(x)| = [L:C_L(x)]$$

for each π-element $x\in L-L_0$. Hence

$$|G_0| = |G| - \sum_x |\mathscr{S}_\pi^G(x)|$$

where the summation is taken over representatives of the conjugacy classes of π-elements in $L-L_0$, and

$$\begin{aligned}|G_0| &= |G| - \sum_x [G:C_L(x)]\\ &= [G:L](|L| - \sum_x [L:C_L(x)])\\ &= [G:L](|L| - \sum_x |\mathscr{S}_\pi^L(x)|)\\ &= [G:L]\cdot|L_0|.\end{aligned}$$

Hence $[G:G_0]=[L:L_0]$, as we require.

Exercises

1. Prove Lemma 2.
2. Show that, in the case that π consists of all primes, then Dade's isometry reduces to the ordinary induction of class functions.
3. Show that the additional assumption in the hypothesis of Theorem 6 may be weakened to the assumption that H contains a Sylow p-subgroup of G for each $p\in\pi$ and that no element of A is conjugate to an element of $H-A$.

4. Assume that, in the hypotheses of Theorem 7, the subgroups $I(x)$ are trivial for all π-elements $x \in L - L_0$. Show that the proof reduces to Suzuki's method of exceptional characters.
5. Obtain analogues of Theorem 7 corresponding to the isometries constructed in Theorem 6 and in Exercise 3.
6. Deduce from Theorem 7 that if P is a Sylow p-subgroup of a group G and any two elements of P which are conjugate in G are already conjugate in P, then G has a normal p-complement.
   [Hint. Use induction on $|G|$.]
7. (Reynolds (1967).) Define *π-induction* as follows.

   Let G be a finite group, H a subgroup of G, and π a set of primes. Let $\mathscr{C}_{G,\pi}$ and $\mathscr{C}_{H,\pi}$ denote the spaces of complex-valued class functions on G and H, respectively, which are constant on π-sections. Show that there is a unique linear map $\theta \to \theta^{G,\pi}$ from $\mathscr{C}_{H,\pi}$ to $\mathscr{C}_{G,\pi}$, called *π-induction,* which is the adjoint of the restriction map, and that a formula for π-induction is given by

$$\theta^{G,\pi}(g) = \frac{[G:H]}{|\mathscr{S}_\pi(g_\pi)|} \sum_h \theta(h)$$

   where the summation is carried over $h \in H \cap \mathscr{S}_\pi(g_\pi)$.
   [Hint. Consider a basis of $\mathscr{C}_{H,\pi}$ given by the characteristic functions of π-sections.]
8. Assume the situation of Exercise 7 and suppose that D is a union of π-sections of H such that the following conditions hold:

   (i) π-elements of D which are conjugate in G are conjugate in H, and
   (ii) if a is a π-element of D, then $C_G(a) = C_H(a) \cdot I(a)$ where $I(a)$ is a normal Hall π'-subgroup of $C_G(a)$.

   Show that π-induction affords a linear isometry from the subspace of $\mathscr{C}_{H,\pi}$ consisting of functions vanishing outside D to $\mathscr{C}_{G,\pi}$ and that

$$\theta^{G,\pi}(g) = \begin{cases} \theta(a) & \text{if } g \in \mathscr{S}_\pi(a) \text{ for } a \in D, \\ 0 & \text{otherwise.} \end{cases}$$

   Show also that $\theta^{G,\pi}$ is a generalised character if θ is, and obtain an analogue also of Theorem 3(i).
   [Hint. Follow the proof of Theorem 3.]
9. Obtain an analogue to Theorem 7 corresponding to π-induction.

2 Block theory

In this section, we shall introduce some of the ideas from the theory of blocks of characters as developed by Brauer. Our approach is slightly unusual since we wish to describe the subject in the spirit of our study of the character ring of a group, and we shall be interested in those properties of blocks which are relevant to the examination of isometries and exceptional characters. Consequently our definition of a block, while equivalent to the usual definitions, will be nonstandard. It turns out, however, that while we will not establish the equivalence, we will still be able to prove all the properties that we shall require, with the exception of Brauer's second and third main theorems (Theorem 15), which we will state without proof. (For proofs of these theorems in particular, and a survey of modular representation theory in general, we refer, for example, to the books by Curtis and Reiner [CR I, II].)

Let G be a group, $x \in G$, and let p be a fixed prime. Recall from Lemma 5.5 that we may write $g = g_p g_{p'}$ where g_p is a p-element, $g_{p'}$ is a p-regular element (i.e., of order coprime to p), and $[g_p, g_{p'}] = 1$. If $g_p \neq 1$, we say that g is *p-singular*. (In general, g_p is called the *p-part* of g and $g_{p'}$ the *p′-part* or *p-regular* part.) If G has s conjugacy classes of p-regular elements, we shall label them (in this section) as $\mathscr{C}_1 = \{1\}, \ldots, \mathscr{C}_s$. Let $\chi_1 = 1, \ldots, \chi_r$ be the irreducible complex characters of G and define, for each i,

$$\hat{\chi}_i(g) = \chi(g_{p'}).$$

Note that these are class functions which are constant on p'-sections. Also define

$$\chi_i^*(g) = \begin{cases} p^a \chi_i(g) & \text{if } g \text{ is } p\text{-regular}, \\ 0 & \text{otherwise}, \end{cases}$$

where p^a is the order of a Sylow p-subgroup of G.

Lemma 8. *The functions $\hat{\chi}_i$ and χ_i^* are generalised characters of G.*

Proof. By Brauer's characterisation of characters, it is sufficient to consider the restrictions of $\hat{\chi}_i$ and χ_i^* to an elementary subgroup E of G. If $E = X \times Y$ where X is a p-group and Y is a p'-group, then $\hat{\chi}_i|_E = 1_X \otimes \chi_i|_Y$. Similarly, $\chi_i^*|_E = p^a |X|^{-1} \pi_E \otimes \chi_i|_Y$ where π_E is the regular character of E.

Lemma 9. *The $\mathbb{Z}$-module spanned by the class functions $\hat{\chi}_1, \ldots, \hat{\chi}_r$ is free of rank s.*

Proof. Let $\mathscr{M}$ denote this module. Then $\mathscr{M}$ is certainly free since $\mathscr{M} \subseteq \mathscr{C}\mathscr{h}(G)$. $\mathbb{C} \otimes_{\mathbb{Z}} \mathscr{M}$ is clearly isomorphic to the space of class functions which vanish on the p-singular elements of G: this has dimension s and thus $\mathscr{M}$ has rank at least s.

Let $\mathscr{M}^*$ be the $\mathbb{Z}$-module spanned by $\chi_1^*, \ldots, \chi_r^*$. Then $\mathscr{M}$ and $\mathscr{M}^*$ are isomorphic. Let $\mathscr{M}_p$ denote the module of generalised characters vanishing on all p-singular elements. Then $\mathscr{M}_p$ contains $\mathscr{M}^*$ and is a direct summand of $\mathscr{Ch}(G)$ of rank s by Proposition 3.1 and Theorem 3.7. Hence $\mathscr{M}$ has rank s.

Let $\zeta_1, \ldots, \zeta_r$ be the restrictions of the irreducible characters of G to the p-regular elements. Then the $\mathbb{Z}$-module $\mathscr{Ch}(G;p)$ that they generate is naturally isomorphic to the module $\mathscr{M}$ of the above lemma and any basis for $\mathscr{Ch}(G;p)$ is called a *basic set*. Since $\zeta_1, \ldots, \zeta_r$ span the space of complex-valued class functions on the p-regular elements of G, the members of any basic set are linearly independent over $\mathbb{C}$.

If $\{\varphi_1, \ldots, \varphi_s\}$ is a basic set, then there are integers $\{d_{ij}\}$ such that

$$\zeta_i = \sum_{j=1}^{s} d_{ij}\varphi_j;$$

these integers are called the *decomposition numbers* with respect to the basic set and the $r \times s$ matrix $D = (d_{ij})$ is called the *decomposition matrix*. Let $C = (c_{ij})$ be the $s \times s$ matrix $D^{\mathrm{T}}D$ (where D^{T} denotes the transpose of D); then C is the *Cartan matrix* and the integers c_{ij} the *Cartan invariants*, all with respect to the given basic set.

Let B be a nonempty subset of $\{\chi_1, \ldots, \chi_r\}$. Then B is said to be a (p-)*block* of G if, whenever $a_1, \ldots, a_r$ are integers such that $\sum a_i\zeta_i = 0$,

$$\sum_{\chi_i \in B} a_i\zeta_i = 0$$

and B is minimal with respect to this property. We observe that this condition actually partitions the set of irreducible characters of G. In the first place,

$$\sum_{\chi_i \notin B} a_i\zeta_i = 0.$$

Then, if B and B' were distinct blocks and $\sum a_i\zeta_i = 0$,

$$\sum_{\chi_i \in B'} a_i\zeta_i = 0,$$

would be a 'new' relation $\sum b_i\zeta_i = 0$ where

$$b_i = \begin{cases} a_i & \text{if } \chi_i \in B'. \\ 0 & \text{otherwise,} \end{cases}$$

and the deduction that

$$\sum_{\chi_i \in B'} b_i\zeta_i = \sum_{\chi_i \in B \cap B'} a_i\zeta_i = 0$$

would imply that $B \cap B' = \varnothing$.

This partition of the set of irreducible characters of G immediately leads to a direct sum decomposition of $\mathscr{Ch}(G;p)$ according to blocks. We shall always assume, therefore, that basic sets will be chosen in such a way that they contain a basis for each summand. Thus we can refer to a basic set for a block; a consequence of this choice is that the decomposition and Cartan matrices will be naturally partitioned according to blocks also. Then we may obtain the following refinements of our definition of blocks and of the orthogonality relations.

Proposition 10. *Let B be a block of G and suppose that $a_1, \ldots, a_r$ are complex numbers such that $\sum_i a_i \zeta_i = 0$. Then $\sum' a_i \zeta_i = 0$ where the summation is restricted to the characters in B. In particular, for any p-regular element g and p-singular element h in G,*

$$\sum_{\chi \in B} \chi(g)\chi(h) = 0.$$

Remark. This actually characterises blocks in the sense that any set of characters for which this condition holds is a union of blocks. (See Osima (1955).)

Proof. If $\sum_i a_i \zeta_i = 0$ and $\{\varphi_1, \ldots, \varphi_s\}$ is a basic set, then $\sum_i \sum_j a_i d_{ij} \varphi_j = 0$ and hence, by the linear independence of $\varphi_1, \ldots, \varphi_s$,

$$\sum_i a_i d_{ij} = 0$$

for each $j = 1, \ldots, s$. If we consider only those $\varphi_j \in B$, we have $d_{ij} = 0$ unless $\chi_i \in B$; hence we may restrict the above sums to those i for which $\chi_i \in B$. But now, summing over j, we see that

$$\sum_{\chi_i \in B} a_i \zeta_i = 0.$$

The second statement follows immediately from the orthogonality relations by putting $a_i = \chi_i(h)$.

We shall be especially interested in the block which contains the principal character of G; this is called the *principal block* and will be denoted by $B_0(G)$. It is a well-known fact that the intersection of the kernels of the irreducible characters in $B_0(G)$ is just $O_{p'}(G)$ (see, for example, either Brauer (1964, I) or Feit [F*; IV.4.12]); we shall prove only that part of this result which we shall require later, following an argument due to Robinson.

We begin by defining a relation on the set of irreducible characters of G by

$$\chi \sim \chi' \quad \text{if} \quad (\chi^*, \chi'^*) \neq 0$$

where χ^* and χ'^* are the generalised characters corresponding to χ and χ' as described in Lemma 8, and we extend this to an equivalence relation $\approx$ by transitivity.

Lemma 11. *Let $\mathscr{B}$ be an equivalence class of irreducible characters under the relation $\approx$. Then $\mathscr{B}$ is a union of blocks.*

Remark. Robinson has proved that $\mathscr{B}$ is a single block so that this equivalence relation actually characterises blocks, but we shall not need this stronger assertion.

Proof. Suppose, in our general notation, that $\sum a_i\zeta_i = 0$. Then $\sum a_i\chi_i^* = 0$, and it will be sufficient to show that $\sum' a_i\chi_i^* = 0$, where the summation is restricted to those i for which $\chi_i \in \mathscr{B}$. Now, if $\sum'' a_i\chi_i^*$ denotes the summation for characters $\chi_i \notin \mathscr{B}$ and $\chi_j \in \mathscr{B}$, certainly

$$(\textstyle\sum'' a_i\chi_i^*, \chi_j) = 0$$

and hence

$$(\textstyle\sum' a_i\chi_i^*, \chi_j) = 0,$$

while, if $\chi_j \notin \mathscr{B}$,

$$(\textstyle\sum' a_i\chi_i^*, \chi_j) = p^{-a}(\textstyle\sum' a_i\chi_i^*, \chi_j^*) = 0$$

by definition. Thus $\sum' a_i\chi_i^* = 0$ as we required.

Theorem 12. *Let G be a group and N a normal p'-subgroup of G. Then $N \subseteq \ker\chi$ for every irreducible character $\chi \in B_0(G)$, and $B_0(G/N)$ coincides with $B_0(G)$ under inflation.*

Proof. In view of Lemma 11, it will be sufficient for the first statement to show that $N \subseteq \ker\chi$ whenever $1_G \approx \chi$. So suppose that

$$1_G = \chi_1 \sim \chi_2 \sim \cdots \sim \chi_n = \chi$$

and assume, for some i, that $N \subseteq \ker\chi_i$ but $N \nsubseteq \ker\chi_{i+1}$. By Clifford's theorem, the distinct irreducible constituents of $\bar{\chi}_{i+1}|_N$ are conjugate so that none is trivial. Consequently, since

$$(\chi_i\bar{\chi}_{i+1})|_N = \chi_i(1)\cdot\bar{\chi}_{i+1}|_N,$$

no constituent of $\chi_i\bar{\chi}_{i+1}$ has N in its kernel. Let ψ be an irreducible constituent of $\chi_i\bar{\chi}_{i+1}$. Since $N \subseteq O_{p'}(G)$, the set $G_{p'}$ of p-regular elements of G forms a union of subgroups of G containing N; hence, as $(1_H, \psi|_H) = 0$ for any subgroup H of G containing N, it follows that

$$\sum_{g \in G_{p'}} \psi(g) = 0.$$

(See Exercise 20 of Section 2.3.) Then $(\chi_i^*, \chi_{i+1}^*) = 0$, which is a contradiction.

Now the irreducible characters of $B_0(G)$ may be considered as characters of G/N and so form a union of blocks of G/N by Lemma 11; conversely, the same is true of characters in $B_0(G/N)$ inflated to characters of G. Thus the principal blocks coincide.

Corollary 13. *Let G be a group with a normal p-complement N. Then $B_0(G)$ consists precisely of those irreducible characters having N in their kernel. In particular, the rank of the corresponding submodule of $\mathscr{C}\mathscr{k}(G; p)$ is 1 and the Cartan invariant corresponding to the principal block is $[G:N]$.*

Proof. G/N has only one conjugacy classes of p-regular elements, so the result is obvious.

We now turn to the major results of Brauer which will relate the characters of G with the characters of certain subgroups in such a way that block relations hold.

Let x be a p-element of G, let $\{\tilde{\chi}_j\}$ be the irreducible characters of $C_G(x)$, and let $\{\tilde{\varphi}_j\}$ be a basic set for $C(x)$. Then there are nonnegative integers $\{a_{ij}\}$ such that $\chi_i|_{C(x)} = \sum_j a_{ij}\tilde{\chi}_j$. Since $x \in Z(C(x))$, there is a root of unity ω_j such that $\tilde{\chi}_j(xy) = \omega_j\tilde{\chi}_j(y)$ for each p-regular element $y \in C(x)$. Hence there are algebraic integers $\{d_{ij}^{(x)}\}$ such that

$$\chi_i(xy) = \sum_j d_{ij}^{(x)}\tilde{\varphi}_j(y)$$

for each p-regular element y of $C(x)$. These algebraic integers $\{d_{ij}^{(x)}\}$ are called the *generalised decomposition numbers* associated with x. From their construction, we see that they lie in the field of p^bth roots of unity where x has order p^b; by the linear independence of a basic set, they are uniquely determined.

Proposition 14. *Assume the situation above. Let $D^{(x)}$ be the matrix of generalised decomposition numbers associated with x and let $\tilde{C}$ be the Cartan matrix of $C(x)$. Then $(D^{(x)})^{\mathrm{T}}\overline{D^{(x)}} = \tilde{C}$.*

Proof. Let $H = C_G(x)$ and let $\tilde{\mathscr{C}}_1, \ldots, \tilde{\mathscr{C}}_t$ be the conjugacy classes of p-regular elements of H. If $\tilde{y}_k \in \tilde{\mathscr{C}}_k$, then $C_G(x\tilde{y}_k) = C_H(\tilde{y}_k)$. Let $X^{(x)}$ be the $r \times t$ matrix $(\chi_i(x\tilde{y}_j))$, $\tilde{Y}$ the $t \times t$ matrix $(|C_H(\tilde{y}_j)|\delta_{ij})$, $\tilde{\Phi}$ the $t \times t$ matrix $(\tilde{\varphi}_i(\tilde{y}_j))$ and $\tilde{D}$ the decomposition matrix of H with respect to the basic set $\{\tilde{\varphi}_j\}$. Then the orthogonality relations for H imply that

$$\tilde{Y} = (\tilde{D}\tilde{\Phi})^{\mathrm{T}}(\overline{\tilde{D}\tilde{\Phi}}) = (\tilde{\Phi}^{\mathrm{T}}\tilde{D}^{\mathrm{T}})(\tilde{D}\bar{\tilde{\Phi}}) = \tilde{\Phi}^{\mathrm{T}}\tilde{C}\bar{\tilde{\Phi}}.$$

On the other hand, from the orthogonality relations for G, we can derive the matrix relations

$$(X^{(x)})^{\mathrm{T}}(\bar{X}^{(x)}) = (|C_G(x\tilde{y}_j)|\delta_{ij}) = (|C_H(\tilde{y}_j)|\delta_{ij}) = \tilde{Y}$$

while

$$(X^{(x)})^{\mathrm{T}}(\bar{X}^{(x)}) = (D^{(x)}\tilde{\Phi})^{\mathrm{T}}(\overline{D^{(x)}\tilde{\Phi}}) = \tilde{\Phi}^{\mathrm{T}}(D^{(x)})^{\mathrm{T}}\overline{D^{(x)}}\,\bar{\tilde{\Phi}}.$$

Since the members of a basic set are linearly independent, $\tilde{\Phi}$ is nonsingular and hence

$$(D^{(x)})^{\mathrm{T}}\overline{D^{(x)}} = \tilde{C}.$$

Now, in a way which we shall not describe, Brauer showed that blocks of G *dominate* certain blocks of $C_G(x)$; in particular, a block of G may dominate any number of blocks of $C_G(x)$, possibly none, but a block of $C_G(x)$ is always dominated by exactly one block of G.

Theorem 15. *Assume the situation described above, with the given notation. Then the following hold.*

(i) *The generalised decomposition number $d_{ij}^{(x)}$ is zero unless $\tilde{\varphi}_j$ belongs to a block of $C_G(x)$ dominated by the block of G which contains the character χ_i.*

(ii) *$B_0(G)$ dominates $B_0(C_G(x))$ and no other.*

Parts (i) and (ii) are known as *Brauer's Second and Third Main Theorems* respectively. These results are highly nontrivial, and their proofs lie at the heart of modular representation theory; it does not seem possible to obtain them directly by arguments based only on the study of ordinary characters, so we shall not prove them. Accepting this point however, we shall give two important applications. The first is a refinement of the property that we used to define blocks.

Proposition 16. *Let B be a block of G and let x be a p-element of G. If $\sum_i a_i\chi_i(xy) = 0$ for all p-regular elements y of $C(x)$ with $a_i \in \mathbb{C}$, then $\sum_i' a_i\chi_i(xy) = 0$ where the summation is carried over the irreducible characters in B.*

Proof. Suppose that $\sum_i a_i\chi_i(xy) = 0$. With the notation previously established, we have

$$\chi_i(xy) = \sum_j d_{ij}^{(x)}\tilde{\varphi}_j(y);$$

hence

$$\sum_{i,j} a_i d_{ij}^{(x)}\tilde{\varphi}_j(y) = 0$$

for all p-regular elements y of $C(x)$. Since the $\{\tilde{\varphi}_j\}$ form a basic set, their linear independence implies that

$$\sum_{i=1}^{r} a_i d_{ij}^{(x)} = 0$$

for each j. Now, if B is a block of G, summing over those blocks of $C(x)$ dominated by B, we see that

$$\begin{aligned}\sum_{\chi_i \in B} a_i \chi_i(xy) &= \sum_{\chi_i \in B} a_i \sum_j d_{ij}^{(x)} \tilde{\varphi}_j(y) \\ &= \sum_{i=1}^{r} \sum_j a_i d_{ij}^{(x)} \tilde{\varphi}_j(y) = 0.\end{aligned}$$

As a consequence of this proposition, we may refine the formula used when elements which are not strongly real are present. (Recall the discussion of Section 2.9.)

Proposition 17. *Let u and v be involutions in a group G and suppose that x is a nonidentity p-element such that no element whose p-part is conjugate to x is strongly real. Then*

$$\sum{}' \frac{\chi(u)\chi(v)\chi(xy)}{\chi(1)} = 0$$

for all p-regular elements $y \in C(x)$, where the summation is carried over the irreducible characters in any p-block of G.

Proof. The element xy cannot be expressed as the product of conjugates of u and v and so

$$\sum_{\chi} \frac{\chi(u)\chi(v)}{\chi(1)} \overline{\chi(xy)} = 0$$

where the summation is carried over all irreducible characters of G. By Proposition 16, this summation may be restricted to any block and we take the complex conjugate, noting that $\chi(u)$ and $\chi(v)$ are real.

In practice, we shall perform the calculation with respect to the principal block, and in the particular circumstances of the next section, we shall be examining the following special situation (with $p = 2$).

Proposition 18. *Let G be a group in which the centraliser of every nonidentity p-element is p-nilpotent. Then an irreducible character of G which is not zero on all p-singular elements lies in the principal block if and only if it is constant on nontrivial p-sections.*

Proof. Let x be a nonidentity p-element. Then $\tilde{\varphi} = 1_{O_{p'}(C(x))}$ forms a basic set for $B_0(C(x))$ by Corollary 13. If $\chi \in B_0(G)$, then there is a generalised decomposition number $d_\chi^{(x)}$ such that

$$\chi(xy) = d_\chi^{(x)}\tilde{\varphi}(y) = d_\chi^{(x)}$$

for all p-regular elements $y \in C(x)$ by Theorem 15. So χ is constant on p-sections.

Conversely, if χ is an irreducible character of G which is constant on p-sections but not zero on all, pick a nonidentity p-element x such that $\chi(x) \neq 0$. Then the formula

$$\chi(xy) = \chi(x)\tilde{\varphi}(y)$$

determines the corresponding generalised decomposition number by uniqueness, forcing $\chi \in B_0(G)$.

We shall complete this section with a discussion of the so-called blocks of defect zero. As we shall see, the conditions stated in the following result imply that the character in question forms a block by itself – the term 'defect zero' refers to the fact that the degree is divisible by the highest possible power of p and that, in particular, if p divides $|G|$, then no such character can lie in $B_0(G)$. When we refer to blocks of defect zero in the applications of these methods in the next section, it is not strictly necessary; careful examination of the arguments will show that the zero values on p-singular elements could be carried through.

Proposition 19. *Let G be a group of order $p^a n$ where $(p, n) = 1$. Then the following conditions for an irreducible character χ are equivalent.*

(i) *p^a divides $\chi(1)$.*
(ii) *χ vanishes on every p-singular element of G.*

Furthermore, $\{\chi\}$ is a block if and only if these conditions hold.

Proof. First assume (ii). Then χ vanishes on every nonidentity element of a Sylow p-subgroup S and (i) follows from the fact that $(\chi|_S, 1_S) \in \mathbb{Z}$.

Conversely, assume (i). Let χ^* be as defined for Lemma 8. Then χ^* is a generalised character of G and $a_i = (\chi^*, \chi_i) \in \mathbb{Z}$ for each irreducible character χ_i of G. Now define

$$\tilde{\chi}(g) = \begin{cases} \chi(g) & \text{if } g \text{ is } p\text{-regular,} \\ 0 & \text{otherwise.} \end{cases}$$

Then $(\tilde{\chi}, \chi_i) = a_i/p^a$. On the other hand, we may compute directly that

$$(\tilde{\chi}, \chi_i) = |G|^{-1} \sum_j |C(g_j)| \chi(g_j) \overline{\chi_i(g_j)}$$

where the summation is carried over representatives of the conjugacy classes of p-regular elements. However, if

$$\omega_j = \frac{|C(g_j)|\chi(g_j)}{\chi(1)},$$

then, by Theorem 2.7, ω_j is an algebraic integer. Since

$$(\tilde{\chi}, \chi_i) = |G|^{-1} \sum_j \chi(1)\omega_j \overline{\chi_i(g_j)},$$

and p^a divides both $|G|$ and $\chi(1)$, it follows that $(\tilde{\chi}, \chi_i)$ is a p-local integer in some algebraic number field. Since $p^a(\tilde{\chi}, \chi_i) \in \mathbb{Z}$, it follows that $(\tilde{\chi}, \chi_i) \in \mathbb{Z}$. Hence $\tilde{\chi}$ is a generalised character. But now

$$\|\tilde{\chi}\|^2 = |G|^{-1} \sum_{g \in G} |\tilde{\chi}(g)|^2 \leqslant |G|^{-1} \sum_{g \in G} |\chi(g)|^2 = \|\chi\|^2 = 1,$$

with equality only if $\tilde{\chi} = \chi$. Since $\|\tilde{\chi}\|^2$ is a positive integer, this must be so and hence χ vanishes on all p-singular elements of G.

Suppose now that condition (ii) holds. If $\sum a_i \zeta_i = 0$, then we have $a_\chi = (\sum a_i \chi_i, \chi) = 0$ and hence the relation $a_\chi \zeta = 0$ where ζ is the restriction of χ to the p-regular elements. Hence $\{\chi\}$ forms a block by definition. Conversely, if $\{\chi\}$ is a block, then $\chi(1)\chi(g) = 0$ for each p-singular element g by Proposition 10 and thus χ vanishes on p-singular elements.

Exercises

1. Show that, in considering relations $\sum_i a_i \zeta_i = 0$ in the definition of blocks, we could have required that the $a_i \in \mathbb{C}$.
2. Let G be a group, p a prime, x a p-element of G, y a p-regular element of $C(x)$ and $z \in O_{p'}(C(x))$. Suppose that $\chi \in B_0(G)$. By using Theorems 12 and 15, show that $\chi(xyz) = \chi(xy)$.
3. Let g, h lie in different p-sections of a group G. Obtain the following refinement of the second part of Proposition 10; show that, for each block B of G,

   $$\sum_{\chi \in B} \chi(g)\overline{\chi(h)} = 0.$$

   [Hint. Use Proposition 16 in place of the definition of blocks.]

3 Groups with Sylow 2-subgroups isomorphic to $Z_{2^n} \times Z_{2^n}$

In the classification of the finite simple groups, a critical stage is reached in one's ability to apply purely group theoretical methods once a group contains an elementary abelian subgroup of order 8. Simple groups in which this does not occur cause a particular problem, and arithmetic arguments, via the application of character theory, come to the fore. In the case of groups with abelian Sylow 2-subgroups, $Z_2 \times Z_2$ occurs as the Sylow 2-subgroup in the groups $\mathrm{PSL}(2, q)$ for $q \equiv 3, 5 \pmod 8$. However, no other rank 2 abelian 2-group occurs as the Sylow 2-subgroup of a simple group; a simple transfer argument shows this if the 2-group is not homocyclic (see Exercise 1), and in this section we shall examine the homocyclic case. This was originally considered by Brauer (1964, II) using purely the block theoretic methods that he developed; in this section, we shall blend the use of his second main theorem with the method of isometries that we developed in Section 1. We shall only show nonsimplicity; it is an easy group theoretic argument to refine the argument to establish solubility.

Theorem 20. *Let G be a group having Sylow 2-subgroups isomorphic to $Z_{2^n} \times Z_{2^n}$, $n \geqslant 2$. Then G is not simple.*

Let S be a Sylow 2-subgroup of G. If $S \subseteq Z(N_G(S))$, then G has a normal 2-complement by Burnside's transfer theorem. (This would necessarily be the case in the nonhomocyclic case.) So we shall assume throughout that $S \nsubseteq Z(N_G(S))$. Then our ultimate aim will be show that G has a nonprincipal linear character in the principal block which contains S in its kernel.

Let $N = N(S)$. For any group X, we shall write $O(X)$ for $O_{2'}(X)$.

Lemma 21. *The following hold.*

(i) *$N/O(N)$ is a Frobenius group of order $3 \cdot 2^{2n}$.*

(ii) *All involutions in G are conjugate and no element of S of order greater than 2 is conjugate to its inverse. Furthermore, any two 2-elements of N which are conjugate in G are already conjugate in N.*

(iii) *$C_G(g)$ is 2-nilpotent for any 2-singular element g of G.*

Proof. Since $|\mathrm{Aut}\,(S)| = 3 \cdot 2^{4n-3}$, $[N(S):C(S)] = 3$, and an element of $N(S) - C(S)$ acts as a fixed-point-free automorphism of order 3 on S. Now $C(S) = S \times O(C(S))$ and (i) follows. Part (ii) follows by Burnside's lemma (Lemma 1.47), and (iii) holds since the 2-part of g has a 2-nilpotent centraliser by Burnside's transfer theorem.

The set $\mathscr{S}^*$ of irreducible characters of N having $O(N)$ in their kernel consists of three linear characters $\varphi_1 = 1$, φ_2 and φ_3 and t characters $\theta_1, \ldots, \theta_t$ of degree 3 where $t = (2^{2n} - 1)/3$. Each of these nonlinear characters is constant on 2-sections. Let $\theta_0 = \varphi_1 + \varphi_2 + \varphi_3$ and let

$$\mathscr{S} = \{\theta_0, \theta_1, \ldots, \theta_t\}.$$

Then a basis for $\mathbb{Z}_0(\mathscr{S})$, as defined in Section 4.1, is given by

$$\{\theta_0 - \theta_1, \theta_1 - \theta_2, \ldots, \theta_1 - \theta_t\}.$$

Notice that each of these generalised characters of N vanishes on all 2-regular elements of N. Let A denote the set of all 2-elements of N; then Hypothesis II of Section 1 holds by Lemma 21 and $\mathbb{Z}_0(\mathscr{S}) \subseteq \mathscr{M}_{\tilde{N}}(\tilde{A})$ in the notation of Theorem 6. We can now apply that theorem.

Proposition 22. *There exists an isometry $\sigma: \mathbb{Z}_0(\mathscr{S}) \to \mathscr{C}\mathscr{h}(G)$ such that $(\mathscr{S}, \sigma)$ is coherent. Furthermore, σ extends to an isometry $\sigma: \mathscr{S}^* \to \mathscr{C}\mathscr{h}(G)$ such that the following hold.*

(i) *For all $x \in S^{\#}$ and y of odd order in $C_G(x)$, $\varphi_i^\sigma(xy) = 1$ and $\theta_j^\sigma(xy) = \theta_j(x)$ for $i = 1, 2, 3$ and $j = 1, \ldots, t$.*
(ii) $\varphi_1^\sigma = 1_G$.
(iii) *For each j, $\theta_j^\sigma(1) - \sum_{i=1}^3 \varphi_i^\sigma(1) = 0$.*
(iv) *$\mathscr{S}^{*\sigma}$ consists, up to sign, of precisely the ordinary irreducible characters in $B_0(G)$.*

Proof. By Theorem 6, we see that the map $\sigma: \mathbb{Z}_0(\mathscr{S}) \to \mathscr{C}\mathscr{h}(G)$ defined by

$$\alpha^\sigma(xy) = \alpha(x), \quad x \in S^{\#}, y \in O(C(x)),$$
$$\alpha^\sigma(g) = 0, \quad \text{if } g \text{ is 2-regular},$$

is a linear isometry. Notice that the characters in $\mathbb{Z}_0(\mathscr{S})$ vanish on all 2-regular elements of N. Thus

$$(1_G, (\theta_0 - \theta_j)^\sigma)_G = 1, \quad j = 1, \ldots, t.$$

Now the set $\{\theta_1, \ldots, \theta_t\}$ is certainly coherent by Proposition 4.2, and the extension of σ is unique as $t \geqslant 5$. Since $\|(\theta_0 - \theta_1)^\sigma\|^2 = 4$ and $\chi_1 = 1_G$ is involved, there are irreducible characters χ_2, χ_3, χ_4 and signs $\varepsilon_2, \varepsilon_3, \varepsilon_4 = \pm 1$ such that

$$(\varphi_1 + \varphi_2 + \varphi_3 - \theta_1)^\sigma = \chi_1 + \varepsilon_2\chi_2 + \varepsilon_3\chi_3 - \varepsilon_4\chi_4. \tag{3.1}$$

Now

$$((\varphi_1 + \varphi_2 + \varphi_3 - \theta_1)^\sigma, (\theta_1 - \theta_j)^\sigma)_G = -1 \quad \text{for } j \geqslant 2.$$

Since $t \geqslant 5$, of the characters $\theta_1^\sigma, \ldots, \theta_t^\sigma$, only θ_1^σ can be involved in $(\varphi_1 + \varphi_2 + \varphi_3 - \theta_1)^\sigma$, and then with multiplicity -1. So we can take

$$\varepsilon_4\chi_4 = \theta_1^\sigma \tag{3.2}$$

and, if we define

$$\theta_0^\sigma = \chi_1 + \varepsilon_2\chi_2 + \varepsilon_3\chi_3,$$

we obtain the coherence of $(\mathscr{S}, \sigma)$. Now, if we define

$$\varphi_i^\sigma = \varepsilon_i\chi_i$$

for $i = 1, 2, 3$ and put $\varepsilon_1 = 1$, we obtain the extension of the isometry to $\mathscr{S}^*$, and (ii) and (iii) follow immediately.

We shall now prove (i) and (iv) together; these will establish the assertion that $\mathscr{S}^{*\sigma}$ constitutes $B_0(G)$ and will determine values on 2-sections. (Recall from Proposition 18 that, since 2-elements have 2-nilpotent centralisers, characters in $B_0(G)$ are constant on nontrivial 2-sections.)

First observe that no θ_j^σ vanishes on every element of even order; otherwise θ_1^σ, say, would lie in a block of defect zero and $2^{2n}|\theta_1^\sigma(1)$. But then $2^{2n}|\theta_j^\sigma(1)$ for all j and hence *every* θ_j^σ would vanish on all 2-singular elements, contrary to the fact that

$$\theta_1^\sigma(xy) - \theta_j^\sigma(xy) = \theta_1(x) - \theta_j(x)$$

for all $x \in S^\#$ and $y \in O(C(x))$.

Indeed, for $x \in S^\#$ and $y \in O(C(x))$, we may write

$$\theta_j^\sigma(xy) = \theta_j(x) + c(x, y) \tag{3.3}$$

where $c(x, y)$ is independent of j. For given j, we may choose x_1, y_1 such that $\theta_j^\sigma(x_1 y_1) \neq 0$. Choose $k \neq j$ such that $\theta_k^\sigma(x_1 y_1) \neq \theta_j^\sigma(x_1 y_1)$; this is possible since we may even find a faithful character θ_k of $N/O(N)$ such that $\theta_k(x_1) \neq \theta_j(x_1)$. Now let $\{\tilde{\varphi}_l\}$ be a basic set for $C_G(x_1)$ chosen to respect blocks, with $\tilde{\varphi}_1 = 1$ for the principal block. Then the generalised decomposition numbers are algebraic integers $\{\tilde{d}_{jl}\}$ and $\{\tilde{d}_{kl}\}$ such that

$$\theta_j^\sigma(x_1 y) = \sum \tilde{d}_{jl} \tilde{\varphi}_l(y)$$

and

$$\theta_k^\sigma(x_1 y) = \sum \tilde{d}_{kl} \tilde{\varphi}_l(y)$$

for all $y \in O(C(x_1))$. So

$$\theta_j^\sigma(x_1 y) - \theta_k^\sigma(x_1 y) = \sum_l (\tilde{d}_{jl} - \tilde{d}_{kl}) \tilde{\varphi}_l(y).$$

On the other hand,

$$\theta_j^\sigma(x_1 y) - \theta_k^\sigma(x_1 y) = \theta_j(x_1) - \theta_k(x_1) \neq 0$$

and, in particular, this expression is independent of y. Thus $\tilde{d}_{jl} = \tilde{d}_{kl}$ for $l \neq 1$ and, taking $y = y_1$, we see that $\tilde{d}_{j1} \neq \tilde{d}_{k1}$. If $\tilde{d}_{k1} \neq 0$, then, by Brauer's second and third main theorems, θ_k^σ belongs to the principal block and

$$\tilde{d}_{jl} = \tilde{d}_{kl} = 0 \quad \text{for } l \neq 1.$$

Hence $\tilde{d}_{j1} \neq 0$ so that $\theta_j^\sigma \in B_0(G)$. This holds for all j so that *all* such characters lie in the principal block.

In particular, θ_j^σ is always constant on 2-sections, and we may write (3.3) as

$$\theta_j^\sigma(xy) = \theta_j(x) + c_x. \tag{3.4}$$

Hence

$$\tilde{d}_{j1} = \theta_j(x) + c_x.$$

Let $\psi_1, \ldots, \psi_u$ be the characters of $B_0(G)$ distinct from $\chi_1, \{\pm\theta_j^\sigma\}$. The Cartan invariant of $B_0(C(x))$ is 2^{2n}; hence, by Corollary 13 and

Proposition 14,

$$1+\sum_{i=1}^{u}|\psi_i(x)|^2+\sum_{j=1}^{t}|\theta_j^\sigma(x)|^2=2^{2n},$$

or

$$1+\sum_{i=1}^{u}|\psi_i(x)|^2+\sum_{j=1}^{t}|\theta_j(x)|^2+c_x\sum_{j=1}^{t}\overline{\theta_j(x)}+\bar{c}_x\sum_{j=1}^{t}\theta_j(x)+t|c_x|^2=2^{2n}.$$

On the other hand, by considering N, or more precisely $N/O(N)$, we have relations

$$3+\sum_{j=1}^{t}|\theta_j(x)|^2=2^{2n}$$

and

$$3+3\sum_{j=1}^{t}\theta_j(x)=0.$$

Hence

$$\sum_{i=1}^{u}|\psi_i(x)|^2-c_x-\bar{c}_x+t|c_x|^2=2. \tag{3.5}$$

Take x to be an involution z. Then $c_z\in\mathbb{Z}$ so that $c_z=0$ as $t\geqslant 5$. Now, for all $y\in O(C(z))$, (3.1) and (3.2) yield that

$$\varepsilon_2\chi_2(zy)+\varepsilon_3\chi_3(zy)=2.$$

From (3.5), $|\varphi_i(z)|=0$ or 1 for each i. Now a similar argument to that used between θ_j^σ and θ_k^σ forces χ_2 and χ_3 into $B_0(G)$ and

$$\varepsilon_2\chi_2(zy)=\varepsilon_3\chi_3(zy)=1.$$

Putting $\varphi_2^\sigma=\varepsilon_2\chi_2$ and $\varphi_3^\sigma=\varepsilon_3\chi_3$ again, φ_2^σ and φ_3^σ are constant on 2-sections. For $x\in S^{\#}$, put

$$\varphi_2^\sigma(x)=1+a_x \quad \text{and} \quad \varphi_3^\sigma(x)=1+b_x.$$

Then

$$a_x+b_x=c_x.$$

Putting $\psi_1=\chi_2$ and $\psi_2=\chi_3$, (3.5) becomes

$$\sum_{i=3}^{u}|\psi_i(x)|^2+1+|a_x|^2+1+|b_x|^2+a_x+\bar{a}_x+b_x+\bar{b}_x$$
$$-c_x-\bar{c}_x+t|c_x|^2=2.$$

Hence

$$a_x=b_x=c_x=\psi_i(x)=0, \quad i=3,\ldots,u.$$

Thus any further characters in $B_0(G)$ vanish on all 2-singular elements, which is impossible by Proposition 19. So (i) and (iv) hold.

Corollary 23. *The following congruences modulo 2^{2n} hold:*

$$\varphi_i^\sigma(1)\equiv 1 \quad \textit{and} \quad \theta_j^\sigma(1)\equiv 3.$$

Proof. These follow immediately using the fact that $(1_S, \chi_S) \in \mathbb{Z}$ for any character χ of G.

We may now complete the proof of Theorem 20. If an involution inverts an element of even order, then it inverts the 2-part; hence, by Lemma 21(ii), no involution can invert an element of order divisible by 4. We shall apply Proposition 17 to restrict the corresponding character theoretic formula to the principal block. (This is the only occasion on which we actually need $n \geqslant 2$, although it was useful to have $t \geqslant 5$.) If z is an involution and u an element of order 4, this yields a formula

$$1 + \frac{1}{\varphi_2^\sigma(1)} + \frac{1}{\varphi_3^\sigma(1)} + \sum_{j=1}^{t} \frac{\theta_j(z)^2 \theta_j(u)}{\theta_j^\sigma(1)} = 0. \tag{3.6}$$

From $N/O(N)$, we see that

$$3 + \sum_{j=1}^{t} \frac{\theta_j(z)^2 \theta_j(u)}{3} = 0;$$

putting $\varphi_2^\sigma(1) = a$ and $\varphi_3^\sigma(1) = b$, the isometry shows that

$$\theta_j^\sigma(1) = 1 + a + b$$

so that the formula (3.6) reduces to a diophantine equation

$$1 + \frac{1}{a} + \frac{1}{b} - \frac{9}{1 + a + b} = 0. \tag{3.7}$$

If $1 + a + b = 3$, then

$$\frac{1}{a} + \frac{1}{b} = 2,$$

forcing $a = b = 1$. Otherwise,

$$|a| \geqslant 15, \quad |b| \geqslant 15 \quad \text{and} \quad |1 + a + b| \geqslant 13$$

from the congruences in Corollary 23 and

$$1 + \frac{1}{a} + \frac{1}{b} - \frac{9}{1 + a + b} \geqslant 1 - \frac{1}{15} - \frac{1}{15} - \frac{9}{13} = \frac{34}{195} > 0,$$

contrary to (3.7). So φ_2^σ and φ_3^σ are linear characters and their kernels contain S since $S \subseteq N' \subseteq G'$.

We make two observations about the proof which we shall examine in a general setting in the next section. The properties that we established for the characters of the principal block of G by using Brauer's second main theorem would have held equally if $G = N$; hence the characters of N with which we started constitute $B_0(N)$. So the isometry σ was actually an isometry from $B_0(N)$ to $B_0(G)$. This we shall show is part of a general phenomenon. The second observation is that we defined our initial isometry on a set of generalised characters which were constant on 2-sections and zero on 2-regular elements; thus we shall want to examine the relationships between these methods and the general theory developed in Section 3.1.

Exercises

1. Show that a group having a Sylow 2-subgroup isomorphic to $Z_{2^m} \times Z_{2^n}$ with $m \neq n$ has a normal 2-complement.
2. Show that the group G of Theorem 20 is soluble.

4 Blocks and exceptional characters

Let G be a group. In this section, rather than work with the field of complex numbers, we shall take a field K which is an algebraic number field and which is a splitting field for G and all its subgroups. We shall restrict our use of the character ring to the usual $\mathbb{Z}$-linear combinations of characters. Since our purpose at this stage is to illustrate the type of results available and, in particular, to put the calculations of the previous section into a more general setting, we will not be exhaustive. The results presented have their origins in the work of Gorenstein and Walter (1962) and Wong (1966) although we shall not necessarily follow them; for a fuller treatment, see Feit [F*; pp. 215–26], and the references contained there. The reader should be warned that precise definitions in this area vary according to source and context, so that our definitions will be tailored to the situations considered.

Let H be a group and D a normal subset of H. Recall the notation introduced in Section 3.1, namely

$$\mathscr{M}_H(D) = \{\theta \in \mathscr{C}\mathscr{h}(H) \mid \theta \text{ vanishes on } H - D\},$$
$$\hat{D} = H - D,$$
$$\hat{\mathscr{M}}_H(D) = \{\theta \in \mathscr{C}\mathscr{h}(H) \mid (\theta, \eta)_H = 0 \text{ for all } \eta \in \mathscr{M}_H(D)\},$$

and define

$$\mathscr{C}_{H,K}(D) = \{\varphi \in \mathscr{C}(H, K) \mid \varphi \text{ vanishes on } H - D\}.$$

Recall also that D is *closed* if $x^n \in D$ whenever $x \in D$ and $\langle x^n \rangle = \langle x \rangle$. In Section 3.1, we showed that, if D is closed and consists of t conjugacy classes, then $\mathscr{M}_H(D)$ is a direct summand of $\mathscr{C}\mathscr{h}(H)$ of rank t and $\hat{\mathscr{M}}_H(D) = \mathscr{M}_H(\hat{D})$.

Now fix a prime p. A normal subset D of H is *complete* if D is a union of p-sections of H. We shall decompose $\mathscr{M}_H(D)$ according to blocks; if B

is a p-block of H, let

$$\mathcal{M}_H(D;B)=\left\{\theta\in\mathcal{M}_H(D)\,\middle|\,\theta=\sum_{\chi\in B}a_\chi\chi\right\}.$$

Proposition 24. *If D is a complete normal subset of H, then*

$$\mathcal{M}_H(D)=\bigoplus\mathcal{M}_H(D;B)$$

where the direct sum is taken over all blocks B of H.

Proof. If $\theta=\sum_i a_i\chi_i\in\mathcal{M}_H(D)$, for each block B put

$$\theta_B=\sum_{\chi_i\in B}a_i\chi_i.$$

Since $H-D$ is a union of p-sections, $\theta_B\in\mathcal{M}_H(D)$ by Proposition 16 (the conclusion of which holds for the trivial p-section consisting of the p-regular elements by the definition of a block), and $\theta=\sum_B\theta_B$.

In order to determine the rank of $\mathcal{M}_H(D;B)$, we must pass to a suitable space of functions. Define

$$\mathscr{C}_{H,K}(D;B)=\left\{\varphi\in\mathscr{C}_{H,K}(D)\,\middle|\,\varphi=\sum_{\chi\in B}c_\chi\chi,c_\chi\in K\right\}.$$

Proposition 25. *Suppose that D is a closed complete normal subset of H. Then* $\dim_K\mathscr{C}_{H,K}(D;B)=\mathbb{Z}$*-rank of* $\mathcal{M}_H(D;B)$.

Proof. By Proposition 24 and Proposition 3.1, $\mathcal{M}_H(D;B)$ is a direct summand of $\mathscr{C}\mathscr{h}(H)$ and a $\mathbb{Z}$-basis is linearly independent over K. Hence

$$\mathrm{rk}_{\mathbb{Z}}\mathcal{M}_H(D;B)\leqslant\dim_K\mathscr{C}_{H,K}(D;B).$$

On the other hand, if D consists of t conjugacy classes of elements, by Theorem 3.7

$$\sum_B\mathrm{rk}_{\mathbb{Z}}\mathcal{M}_H(D;B)=t$$

while

$$\sum_B\dim_K\mathscr{C}_{H,K}(D;B)\leqslant\dim_K\mathscr{C}_{H,K}(D)=t;$$

thus equality holds throughout.

We can now determine the ranks precisely.

Theorem 26. *Let D be a closed complete normal subset of H and let B be a block of H. Let $\{x_1,\ldots,x_t\}$ be a set of representatives of the conjugacy* *classes of p-elements in D and let m_i be the number of members of a basic set for the union of those blocks of $C_H(x_i)$ dominated by B. Then $\mathscr{M}_H(D;B)$ has rank $\sum_i m_i$.*

Proof. By Proposition 25, it suffices to determine $\dim_K \mathscr{C}_{H,K}(D)$. If $\theta \in \mathscr{C}_{H,K}(D;B)$, there exist complex numbers $c_\varphi^{(i)}$ analogous to generalised decomposition numbers such that, for all p-regular elements $y_i \in C_H(x_i)$,

$$\theta(x_i y_i) = \sum_\varphi c_\varphi^{(i)} \varphi(y_i)$$

where the sum is taken over a basic set for the blocks of $C_H(x_i)$ dominated by B. In a natural way, the functions φ that occur induce class functions on H whose span contains θ, and hence

$$\dim_K \mathscr{C}_{H,K}(D;B) \leqslant \sum m_i.$$

On the other hand, suppose that for some block there were strict inequality. Then, summing over all blocks, we see that every member of a complete basic set for each $C_H(x_i)$ appears and we would have the strict inequality

$$\begin{aligned}\dim_K \mathscr{C}_{H,K}(D) &< \sum_i (\text{number of members of a basic set for } C(x_i))\\ &= \sum_i (\text{number of classes of } p\text{-regular elements in } C(x_i))\\ &= \text{number of conjugacy classes in } D\end{aligned}$$

by Lemma 2. But these are clearly equal, a contradiction.

The following definition will generalise the concept of special subsets introduced in Section 3.1 and, with the restriction that $\pi = \{p\}$, generalises the hypotheses considered in Section 1 (though is related to the situation when p-induction can be used).

Definition. Suppose that H is a subgroup of a group G. A normal subset D of p-singular elements of H will be called *special* if the following conditions hold:

(i) if a_1, a_2 are p-elements of D which are conjugate in G, then they are conjugate in H,
(ii) if a is a p-element in D, then $C_G(a) = C_H(a) \cdot I(a)$ where $I(a)$ is a normal p-complement of $C_G(a)$.

In order to pursue the ideas of Section 1 in the spirit of Chapter 4, we shall examine consequences of the following hypothesis.

Hypothesis III. (a) G is a group and p a prime. H is a subgroup of G and D is a complete special normal subset of H (with respect to the prime p).

Let $\tilde{D}$ be the union of p-sections of G which intersect D and put

$$\mathscr{C}^*_{H,K}(D) = \{\theta \in \mathscr{C}_{H,K}(D) \mid \theta \text{ is constant on each } p\text{-section in } D\}.$$

(b) $\mathscr{C}$. is a subspace of $\mathscr{C}^*_{H,K}(D)$ and there is an isometry $\tau: \mathscr{C} \to \mathscr{C}_{G,K}(\tilde{D})$ such that, for each $\theta \in \mathscr{C}$,

(i) θ^τ takes the value $\theta(a)$ on the p-section $\mathscr{S}_p(a)$ for each p-element $a \in D$,
(ii) if $\alpha \in \mathscr{C}(G, K)$ and α is constant on each p-section in $\tilde{D}$, then $(\theta^\tau, \alpha)_G = (\theta, \alpha|_H)_H$, and
(iii) if θ is a generalised character in $\mathscr{C}$, then θ^τ is a generalised character of G.

Theorem 27. *Assume that Hypothesis III holds with* $\mathscr{C} = \mathscr{C}_{H,K}(D, B_0(H))$. *Then*

$$\mathscr{M}_H(D, B_0(H))^\tau \subseteq \mathscr{M}_G(\tilde{D}, B_0(G)).$$

If D is closed, then

(i) *if $D^G \cap H = D$ then there is equality above, and*
(ii) *if $\tilde{D}$ contains every nonidentity p-element of G, then every irreducible character in $B_0(G)$ occurs as a constituent of θ^τ for some $\theta \in \mathscr{M}_H(D, B_0(H))$.*

Remark. The argument used in the proof of Proposition 18 shows that $\mathscr{C}_{H,K}(D, B_0(H)) \subseteq \mathscr{C}^*_{H,K}(D)$ so that it is natural to restrict our attention to principal blocks, and necessary for the second part of the theorem. However, an analogue of the first part will hold under suitable hypotheses for nonprincipal blocks with an appropriate statement about block correspondence. (See, for example, Wong (1966).)

Proof. Let $\theta = \sum_i a_i \psi_i \in \mathscr{M}_H(D, B_0(H))$ and suppose that

$$\theta^\tau = \sum_k c_k \chi_k$$

where the ψ_i are irreducible characters in $B_0(H)$ and the χ_k are irreducible characters of G.

Let x be a p-element in D. Then, for any p-regular element y of $C_G(x)$,

$$\chi_k(xy) = \sum_j d^{(x)}_{kj} \varphi_j(y)$$

where the $\{\varphi_j\}$ form a basic set for $C_G(x)$ chosen to respect blocks and $d^{(x)}_{kj}$ are the associated generalised decomposition numbers. Further, since

$C_G(x)$ is p-nilpotent, we may chose $\varphi_1 = 1$ for $B_0(C_G(x))$. So, since $\theta^\tau(xy) = \theta(x)$, we have

$$\sum_{k,j} c_k d_{kj}^{(x)} \varphi_j(y) = \sum_i a_i \psi_i(x) = \left(\sum_i a_i \psi_i(x) \right) \varphi_1(y).$$

This holds for all p-regular elements $y \in C_G(x)$; hence, by the linear independence of the members of the basic set for $C_G(x)$,

$$\sum_k c_k d_{kj}^{(x)} = 0$$

for all $j \neq 1$.

On the other hand, by Proposition 24,

$$\theta^\tau \in \mathscr{M}_G(\tilde{D}) = \bigoplus \mathscr{M}_G(\tilde{D}; B)$$

where the direct sum is taken over all blocks of G. Let

$$\eta = \sum{}' c_k \chi_k$$

where the sum is taken over all nonprincipal blocks of characters. Then, for x a p-element in D and $y \in I(x)$,

$$\eta(xy) = \sum{}' c_k \chi_k(xy) = \sum_{j \neq 1} \left(\sum_k{}' c_k d_{kj}^{(x)} \right) \varphi_j(y) = 0.$$

Thus η vanishes on $\tilde{D}$ and hence is identically zero.

Suppose now that D is closed. If $D^G \cap H = D$, then $\theta^\tau|_H = \theta$ for all $\theta \in \mathscr{M}_H(D, B_0(H))$. Furthermore, since $\tilde{D} \cap H = D$ also, if $\xi \in \mathscr{M}_G(\tilde{D})$ then $\xi|_H \in \mathscr{M}_H(D)$ so that

$$\mathscr{M}_H(D, B_0(H)) \subseteq \mathscr{M}_G(\tilde{D}, B_0(G))|_H \subseteq \mathscr{M}_H(D).$$

As D is closed, both $\mathscr{M}_H(D, B_0(H))$ and $\mathscr{M}_G(\tilde{D}, B_0(G))$ have rank equal to the number of conjugacy classes of p-elements in D by Theorem 26. Since $\mathscr{M}_H(D, B_0(H))$ is a direct summand of $\mathscr{M}_H(D)$, which is a free $\mathbb{Z}$-module,

$$\mathscr{M}_H(D, B_0(H)) = \mathscr{M}_G(\tilde{D}, B_0(G))|_H.$$

Now, if $\xi \in \mathscr{M}_G(\tilde{D}, B_0(G))$, ξ is constant on p-sections and it follows that $\xi = (\xi|_H)^\tau$; hence

$$\mathscr{M}_H(D, B_0(H))^\tau = \mathscr{M}_G(\tilde{D}, B_0(G)).$$

Suppose now that $\tilde{D}$ contains every nonidentity p-element of G. Let $\chi \in B_0(G)$. Then χ is constant on all p-sections of G by Proposition 18. Assume that $(\chi, \theta^\tau) = 0$ for all $\theta \in \mathscr{M}_H(D, B_0(H))$. Then, by hypothesis, $(\chi|_H, \theta) = 0$ for all such θ. Now, as in the proof of Lemma 3.5,

$$|H| \cdot \chi|_H \in \mathscr{M}_H(D) \oplus \mathscr{M}_H(\hat{D}) = \left(\bigoplus_B \mathscr{M}_H(D; B) \right) \oplus \mathscr{M}_H(\hat{D})$$

where the direct sum is taken over all blocks B of H. This is an orthogonal direct sum so that, in view of the preceding remark, we can restrict it to the nonprincipal blocks, and we may write

$$|H|\cdot\chi|_H = \sum\psi_B + \xi$$

where $\psi_B \in \mathcal{M}_H(D; B)$ and $\xi \in \mathcal{M}_H(\hat{D})$, and the sum is taken over the nonprincipal blocks of H.

Since $\chi \in B_0(G)$, for a p-element $x \in D$ and all p-regular elements $y \in C_H(x)$,

$$\chi(xy) = \chi(x) = \chi(x)\varphi_1^{(H)}(y)$$

where $\varphi_1^{(H)}$ is the member of the basic set of $B_0(C_H(x))$, by Brauer's second main theorem. Pick x so that this expression is nonzero. Then

$$|H|\chi(xy) = \sum\psi_B(xy) + \xi(xy) = \sum\psi_B(xy) = |H|\chi(x)\varphi_1^{(H)}(y) \neq 0$$

for all p-regular elements $y \in C_H(x)$, where the sum is again taken over all nonprincipal blocks of H; this is impossible by Brauer's second and third main theorems and the uniqueness of the generalised decomposition numbers. Hence $(\chi, \theta^\tau) \neq 0$ for some $\theta \in \mathcal{M}_H(D, B_0(H))$.

Theorem 27 explains why, in the proof of Theorem 20, we succeeded in constructing every character in the principal block by, essentially, the methods of exceptional character theory rather than by using modular representation theory, other than for the formal use of Brauer's second and third main theorems.

5 Principal 2-block of groups with dihedral Sylow subgroups

Another situation where the isometries constructed in Section 1 satisfy Hypothesis III is for groups with dihedral Sylow 2-subgroups, and we shall construct their principal blocks.

Theorem 28. *Let G be a group which has a dihedral Sylow 2-subgroup S. Suppose that G does not have a subgroup of index 2. Then the principal 2-block contains the following characters.*

(i) *If $|S| = 4$, then $B_0(G)$ consists of four irreducible characters, $\chi_1 = 1_G$, χ_2, χ_3, χ_4. There are signs $\varepsilon_2, \varepsilon_3, \varepsilon_4 = \pm 1$ such that, if x is an element of even order and g is an element of odd order, then $\chi_i(x) = \varepsilon_i$ and*

$$1 + \sum_{i=2}^{4} \varepsilon_i \chi_i(g) = 0.$$

Also, $\chi_i(1) \equiv \varepsilon_i \pmod 4$.

(ii) *If $|S| = 2^n \geq 8$, let $\varphi_1 = 1_S, \varphi_2, \varphi_3, \varphi_4$ be the linear characters of S and $\psi_1, \ldots, \psi_t$ $(t = 2^{n-2} - 1)$ be the irreducible characters of degree 2. Then $B_0(G)$ consists of characters $\chi_1 = 1_G, \chi_2, \chi_3, \chi_4, \xi_1, \ldots, \xi_t$ and there are signs $\varepsilon, \varepsilon_2, \varepsilon_3, \varepsilon_4 = \pm 1$ such that, if $u \in S^\#$ and v is an element of odd* *order in* $C(u)$, *then*

$$\chi_i(uv) = \varepsilon_i \varphi_i(u) \quad \text{and} \quad \xi_j(uv) = \varepsilon \psi_j(u).$$

If g *is an element of odd order, then all characters* ξ_j *take the same value on* g *and*

$$1 + \sum_{i=2}^{4} \varepsilon_i \varphi_i(u) \chi_i(g) + \varepsilon \psi_j(u) \xi_j(g) = 0.$$

Proof. Let $N = N_G(S)$. An easy transfer argument shows that if $|S| = 4$ then $N/O(N) \cong A_4$ and $S \cdot O(N) = S \times O(N)$, while if $|S| > 4$ then $N = S \times O(N)$; in either case, all involutions are conjugate and the centraliser of any element of $S^{\#}$ is 2-nilpotent.

Suppose first that $|S| = 4$. Let u be an involution in S. Then $O(C_N(u)) = O(N)$, and it is trivial to verify that $B_0(N)$ consists of all characters having $O(N)$ in their kernels. Let $H = N$ and let D be the set of elements of even order in H. Then the isometry σ of Theorem 6 satisfies the conditions of Hypothesis III.

Let $\varphi_1 = 1$, φ_2 and φ_3 be the linear characters in $B_0(N)$ and let ψ be the character of degree 3. Then $\mathscr{M}_N(D, B_0(N))$ is spanned by $(\varphi_1 + \varphi_2 + \varphi_3 - \psi)$. By Theorem 27, $\mathscr{M}_G(\tilde{D}, B_0(G))$ is spanned by $(\varphi_1 + \varphi_2 + \varphi_3 - \psi)^\sigma$ and every character in $B_0(G)$ is a constituent. By Theorem 6, $\chi_1 = 1_G$ is a constituent with multiplicity 1; suppose that the signs $\varepsilon_2, \varepsilon_3, \varepsilon_4 = \pm 1$ are chosen so that

$$(\varphi_1 + \varphi_2 + \varphi_3 - \psi)^\sigma = \chi_1 + \varepsilon_2 \chi_2 + \varepsilon_3 \chi_3 + \varepsilon_4 \chi_4.$$

Any character takes integral values on an involution, and

$$\chi_1(u) + \varepsilon_2 \chi_2(u) + \varepsilon_3 \chi_3(u) + \varepsilon_4 \chi_4(u) = 4;$$

also, since the Cartan invariant for $B_0(C_G(u))$ is $|S|$,

$$\sum_i \chi_i(u)^2 = 4.$$

Hence, since characters in $B_0(G)$ are constant on 2-sections, for any element x of even order, $\chi_i(x) = \varepsilon_i$. The assertion about values on elements of odd order follows from Proposition 10, while the congruence for the degrees arises from the inner product $(\chi_i|_S, 1_S) \in \mathbb{Z}$.

Now suppose that $|S| > 4$. Let S_0 be the cyclic subgroup of S of index 2 and let $D = S_0^{\#}$. Then the conditions of Hypothesis I are satisfied with $N = S$, $\pi = \{2\}$ and $A = D$. Let τ be Dade's isometry as in Theorem 3; then Hypothesis III is satisfied with $\mathscr{C} = \mathscr{C}_{S,K}(D)$ and Theorem 27 applies. Notice that every character of S lies in $B_0(S)$. (Note: we could use Dade's isometry when $|S| = 4$ if $N(S)$ contained a subgroup isomorphic to A_4, but this need not be the case.) Let φ_2 be the nonprincipal irreducible character of S having S_0 in its kernel, put $\psi_0 = \varphi_3 + \varphi_4$, and let $\mathscr{S} = \{\psi_1, \ldots, \psi_t\}$. Then a basis for $\mathscr{M}_S(D)$ is given by

$$\varphi_1 + \varphi_2 - \psi_1, \psi_1 - \psi_0, \psi_1 - \psi_i, \quad i = 2, \ldots, t.$$

The analysis now follows the route established in Chapter 3 in that we must first unravel the images of this basis under τ to seek a natural extension of τ to the whole of $B_0(S)$. By Theorem 27, we can restrict our attention to characters in $B_0(G)$, but then all such characters are constant on 2-sections so that we have the full analogue of Frobenius reciprocity available. Also, we have the fact that $\theta^\tau(1) = 0$ for all $\theta \in \mathscr{M}_S(D)$ to show that if $\|\theta_1^\tau\|^2 = \|\theta_2^\tau\|^2 = 3$ and $(\theta_1^\tau, \theta_2^\tau) = 1$, then θ_1^τ and θ_2^τ cannot have the same three irreducible constituents with just a single difference of sign.

If $|S| = 8$, then we have only $\varphi_1 + \varphi_2 - \psi_1$ and $\psi_1 - \psi_0$ to consider. The above observation enables us to deduce that $B_0(G)$ contains precisely 5 characters and to establish the correspondence under τ with $B_0(S)$.

If $|S| \geqslant 16$, we again start with $(\varphi_1 + \varphi_2 - \psi_1)^\tau$ and $(\psi_1 - \psi_0)^\tau$ and label the constituents so that $\varepsilon\xi_1$ is the common constituent with multiplicity -1 in the first and 1 in the second. If $|S| > 16$, then $t \geqslant 7$ and it is easily seen that $\varepsilon\xi_1$ is the common constituent of $(\varphi_1 + \varphi_2 - \psi_1)^\tau$ and all $(\psi_1 - \psi_j)^\tau$. Then, by Proposition 4.2, $(\mathscr{S}, \tau)$ is coherent with the extension of τ to $\mathscr{S}$ unique and we get the desired correspondence. If $|S| = 16$, then $t = 3$ and, as in the proof of Theorem 3.17, the isometry and inner product with χ_1 alone yield an apparent additional solution, namely

$$(\varphi_1 + \varphi_2 - \psi_1)^\tau = \chi_1 + \varepsilon_2\chi_2 - \varepsilon\xi_1,$$
$$(\psi_1 - \varphi_3 - \varphi_4)^\tau = \varepsilon\xi_1 - \varepsilon_3\chi_3 - \varepsilon_4\chi_4,$$
$$(\psi_1 - \psi_2)^\tau = -\varepsilon_2\chi_2 - \varepsilon_3\chi_3,$$
$$(\psi_1 - \psi_3)^\tau = -\varepsilon_2\chi_2 - \varepsilon_4\chi_4.$$

But from these decompositions, we get

$$\chi_1(1) + 3\varepsilon_2\chi_2(1) = 0,$$

which is absurd.

Now that we have achieved the analogue of a natural induction as we defined it at the end of Section 3.2, we observe that, given the precise analogue of Frobenius reciprocity, the argument of Theorem 3.13 carries over and yields the claimed character values on all 2-elements, and hence on all 2-singular elements; the relations for elements of odd order follow from Proposition 10.

6 Groups with quaternion Sylow 2-subgroups

As a concrete application of the calculations of the previous section, we can now give a short proof of the Brauer–Suzuki theorem for an (ordinary) quaternion Sylow 2-subgroup of order 8, following an argument due to Suzuki (1962a).

Theorem 29. *Let G be a group having a quaternion Sylow 2-subgroup S of order 8, and suppose that G has no nonidentity normal subgroup of odd order. Then $|Z(G)| = 2$.*

Proof. Let $S = \langle x, y | x^4 = 1, y^2 = x^2, y^{-1}xy = x^{-1} \rangle$, and put $z = x^2$. If any subgroup of order 4 in S is conjugate in G to no other, then the focal subgroup theorem (Theorem 1.46) implies first that G has a subgroup of index 2 and then that G has a normal 2-complement. So we can assume that all three subgroups of order 4 in S, and hence all elements of S of order 4, are conjugate in $C_G(z)$.

Let $H = C_G(z)$. Then the elements of H of even order form a closed set of special classes. Also $[N_H(S):C_H(S)]$ is divisible by 3, so that H has no subgroup of index 2. Let φ be an irreducible character of H obtained by inflation from one of the nonprincipal characters χ_i of $H/\langle z \rangle$ given by Theorem 28 and suppose that $\varphi(h) = \varepsilon$ whenever h has order divisible by 4. Then the generalised character $\eta = 1 + \varepsilon\varphi$ takes the values

$$\begin{array}{ll} 1 + \varepsilon\varphi(g) & \text{on elements } g \text{ of odd order,} \\ 1 + \varepsilon\varphi(g) & \text{on elements } zg \text{ of twice odd order,} \\ 2 & \text{on elements of order divisible by 4,} \end{array}$$

and, in particular, $\eta(1) \neq 1$.

The main step is to vary these values in order to obtain an irreducible character ψ of H such that the generalised character

$$\theta = 1_H + \varepsilon\varphi - \varepsilon\psi$$

vanishes on all elements of H of odd order. If this can be achieved, then this generalised character θ will play the same role as the corresponding generalised character θ in the generalised quaternion group case (Theorem 3.14) and the proof that $|Z(G)| = 2$ will be exactly as there.

We shall use Brauer's characterisation of characters. Consider the function ζ on H taking values

$$\begin{array}{ll} 1 + \varepsilon\varphi(g) & \text{on elements } g \text{ of odd order,} \\ -1 - \varepsilon\varphi(g) & \text{on elements } zg \text{ of twice odd order,} \\ 0 & \text{on elements of order divisible by 4.} \end{array}$$

Certainly this is a class function since η is. If we can show that ζ is a generalised character, then $\|\zeta\| = 1$ since clearly $0 < \|\zeta\| < \|\eta\|$, and we may take $\psi = \varepsilon\zeta$. So we need only check that $\zeta|_E \in \mathscr{C}\mathscr{h}(E)$ for each elementary subgroup E of H. To do so, we shall use crucially the fact that $\eta|_E \in \mathscr{C}\mathscr{h}(E)$.

Case 1: E has odd order.
In this case, $\zeta|_E = \eta|_E$ and we are done.

Case 2: E has twice odd order.
Let $E = \langle z \rangle \times F$ and let γ be the irreducible character of E with kernel F. Then $\zeta|_E = \gamma \cdot \eta|_E \in \mathscr{C}\mathscr{h}(E)$.

Case 3: E has $4 \times$ odd order.
All subgroups of order 4 are conjugate, so we may suppose that $E = \langle x \rangle \times F$. Let β be the irreducible character of E having F in its kernel with $\beta(x) = i$. Then we may verify that

$$\zeta|_E = \beta \cdot 1_E + \bar{\beta} \cdot (\varepsilon\varphi)|_E \in \mathscr{C}\mathscr{h}(E).$$

Case 4: E has $8 \times$ odd order.
Then we may suppose that $E = S \times T$ where T is a cyclic group of odd order. Let $\xi = -1 + \varepsilon\varphi$. Then ξ vanishes on all elements of H of order divisible by 4. We shall construct $\zeta|_E$ by considering $\xi|_E$.

If $|T| = t$, let $\tau_1, \ldots, \tau_{t-1}$ be the nonprincipal irreducible characters of E having S in their kernels, let ρ be the character of E obtained by the inflation of the regular character of $E/(\langle z \rangle \times T)$, and let σ be the irreducible character of E of degree 2 having T in its kernel. Then $\{\rho, \rho\tau_1, \ldots, \rho\tau_{t-1}, \sigma\tau_1, \ldots, \sigma\tau_{t-1}\}$ forms a basis for the module of generalised characters of E vanishing on elements of order divisible by 4. Suppose that λ, $\{\mu_j\}$ and $\{v_j\}$ are integers such that

$$\xi|_E = \lambda\rho + \sum_j \mu_j \rho\tau_j + \sum_j v_j \sigma\tau_j.$$

Then

$$\varphi|_E = \varepsilon 1_E + \varepsilon\lambda\rho + \sum_j \varepsilon\mu_j \rho\tau_j + \sum_j \varepsilon v_j \sigma\tau_j.$$

However, $\varphi|_E$ is actually a character. Hence all the coefficients $\varepsilon\mu_j$ and εv_j, as also $\varepsilon\lambda$ and $\varepsilon + \varepsilon\lambda$, are nonnegative and, since $\varphi(z) = \varphi(1)$, we conclude that all the coefficients $\varepsilon_j v_j$ are zero. Now

$$\eta|_E = 2(1_E) + \lambda\rho + \sum_j \mu_j \rho\tau_j$$

and it follows by inspection that

$$\zeta|_E = \sigma + (2\lambda)\sigma + 2\sum_j \mu_j \sigma\tau_j.$$

It is perhaps worth remarking that the study of $\varphi|_E$ in the final analysis above is one of the rare instances in the application of exceptional character theory methods where explicit use is made of the fact that a character (rather than just a generalised character) is under investigation.

Exercise

1. Verify that the proof of Theorem 29 may be completed by repeating the arguments of Section 3.3.

7 Some further isometries

In this section, we shall discuss some of the work of Reynolds (1968) and Robinson (1985) and consider the following hypothesis which generalises situations which we have examined previously; specifically, the condition that centralisers have normal π-complements has been dropped.

Hypothesis IV. *G is a finite group, H is a subgroup of G, π is a set of primes, and D is a union of π-sections of H such that*

(i) *any two π-elements in D which are conjugate in G are already conjugate in H, and*

(ii) $C_G(a) = C_H(a)O_{\pi'}(C_G(a))$ *for each π-element $a \in D$.*

We will extend the notation of Section 4; throughout this section, $\tilde{D}$ will denote the union of those π-sections of G which meet D. Our target will be to construct an isometry from a suitable subspace of $\mathscr{C}_H(D)$ to $\mathscr{C}_G(\tilde{D})$ where $\mathscr{C}_H(D)$ is the space of complex-valued class functions on H which vanish outside D and $\mathscr{C}_G(\tilde{D})$ is similarly defined. Furthermore, we will want the isometry to carry generalised characters to generalised characters. As usual, it is not difficult to construct a function which takes sensible values and show that it is an isometry; the hard part will be to show that generalised characters are mapped to generalised characters. In the case of a single prime we shall use block theory and, in particular, some of the ideas from Section 4; for the general case we shall make use of Brauer's characterisation of characters to obtain a reduction to the position where π consists of a single prime (though its essential role is partially masked by the inductive hypothesis).

We shall start by considering a group G satisfying Hypothesis IV with $\pi = \{p\}$. Although what we do is motivated by the results of Section 4, we shall not restrict our attention in this case to functions which are constant on p-sections, but rather seek to exploit properties of the principal block.

Let

$$\mathscr{C}_H(D)^0 = \left\{ \varphi \in \mathscr{C}_H(D) \,\middle|\, \varphi = \sum_{\chi \in B_0(H)} c_\chi \chi, c_\chi \in \mathbb{C} \right\}$$

and define $\mathscr{C}_G(\tilde{D})^0$ analogously. Notice that, if $u \in \tilde{D}$, then its p-part u_p is conjugate to an element of D and we may suppose that $u_p \in D$ for the purpose of defining class functions on G.

Lemma 30. *Let $\varphi \in \mathscr{C}_H(D)^0$. If $u \in \tilde{D}$ and $u_p \in D$, suppose that $u_{p'} = vw$ where v is a p-regular element of $C_H(u_p)$ and $w \in O_{p'}(C_G(u_p))$. Then there is a well-defined class function $\varphi^* \in \mathscr{C}_G(\tilde{D})^0$ such that $\varphi^*(u) = \varphi(u_p v)$ whenever $u \in \tilde{D}$ and $u_p \in D$.*

Proof. Suppose that we also have $u_{p'} = v'w'$, where v' is a p-regular element of $C_H(u_p)$ and $w' \in O_{p'}(C_G(u_p))$. Then

$$v^{-1}v' \in C_H(u_p) \cap O_{p'}(C_G(u_p)) = O_{p'}(C_H(u_p)). \tag{7.1}$$

If ψ is an irreducible character in $B_0(H)$, then $\psi(u_p v) = \psi(u_p v')$ by Theorems 12 and 15 (as in Exercise 2 of Section 2). So we may define a function $\varphi^* \in \mathscr{C}_G(\tilde{D})$ by putting

$$\varphi^*(u) = \varphi(u_p v)$$

whenever $u \in \tilde{D}$ and $u_p \in D$; it remains only to show that $\varphi^* \in \mathscr{C}_G(\tilde{D})^0$.

For each block B of G, write $\varphi_B^* = \sum_{\chi \in B}(\varphi^*, \chi)\chi$; then

$$\varphi^* = \sum_B \varphi_B^*.$$

By Proposition 16, or the definition of blocks for p-regular elements, each function φ_B^* vanishes on every p-section disjoint from $\tilde{D}$. Now suppose that $u \in \tilde{D}$ with $u_p \in D$. In view of Hypothesis IV(ii), (7.1) and Theorem 12, $B_0(C_H(u_p))$ may be identified with $B_0(C_G(u_p))$ by inflation; furthermore, taking a basic set $\{\tilde{\varphi}_i\}$ for this principal block, by Brauer's second main theorem there are complex numbers d_i such that

$$\varphi^*(u) = \varphi(u_p v) = \sum_i d_i \tilde{\varphi}_i(v)$$

and $\tilde{\varphi}_i(v) = \tilde{\varphi}_i(v')$ whenever $u_{p'} = vw = v'w'$ where v, v' are p-regular elements of $C_H(u_p)$ and $w, w' \in O_{p'}(C_G(u_p))$. Hence the support of φ^* lies in $B_0(C_G(u_p))$ when restricted to $\mathscr{S}_p(u_p)$, and φ_B^* vanishes on $\tilde{D}$ whenever $B \neq B_0(G)$. Thus $\varphi^* \in \mathscr{C}_G(\tilde{D})^0$.

Let σ denote the map from $\mathscr{C}_H(D)^0$ to $\mathscr{C}_G(\tilde{D})^0$ defined by Lemma 30.

Lemma 31. *σ is an isometry.*

Proof. Clearly σ is a linear map so it is sufficient to show that $\|\varphi^*\|_G = \|\varphi\|_H$ for each $\varphi \in \mathscr{C}_H(D)^0$. Thus, in view of Hypothesis IV(i), it suffices to show that, for each p-element $a \in D$,

$$|G|^{-1}\sum|\varphi^*(y)|^2 = |H|^{-1}\sum|\varphi(z)|^2,$$

where the first sum is taken over all elements $y \in \mathscr{S}_\pi^G(a)$ and the second sum over the corresponding elements $z \in \mathscr{S}_\pi^H(a)$. Now

$$|G|^{-1}\sum_y |\varphi^*(y)|^2 = |C_G(a)|^{-1}\sum_b |\varphi^*(ab)|^2 \tag{7.2}$$

where the second sum is taken over all p'-elements $b \in C_G(a)$. The p'-elements of $C_G(a)$ form a union of cosets of $O_{p'}(C_G(a))$ having representatives $b_1, \ldots, b_n$ in $C_H(a)$; in view of (7.1), the right-hand side of (7.2) becomes

$$[C_G(a):O_{p'}(C_G(a))]^{-1}\sum_i |\varphi^*(ab_i)|^2$$

which reduces to

$$|C_H(a)|^{-1}\sum |\varphi(ab)|^2$$

where the sum is taken over all p'-elements $b \in C_H(a)$, and this is equal to the required sum $|H|^{-1}\sum|\varphi(z)|^2$.

The map that we have thus constructed is *Reynolds' isometry*. We still need to show that it carries generalised characters to generalised characters, and we shall do this by identifying σ with another map for which this assertion is obvious.

Theorem 32. *Assume Hypothesis IV with* $\pi = \{p\}$. *Then the isometry* σ *defined in Lemmas* 30 *and* 31 *is identical to the map*

$$\varphi \rightarrow (\varphi^G)_0 = \sum_{\chi \in B_0} (\varphi^G, \chi)\chi$$

where $B_0 = B_0(G)$. *If S is any subring of* $\mathbb{C}$ *containing* $\mathbb{Z}$, *then* σ *maps S-linear combinations of characters of H which lie in* $\mathscr{C}_H(D)^0$ *to S-linear combinations of characters of G and, in particular, generalised characters to generalised characters.*

Proof. By an immediate extension of Proposition 24 to class functions, putting $\mathscr{C}_G(\tilde{D}; B) = \mathbb{C} \otimes \mathscr{M}_G(\tilde{D}; B)$, we have an orthogonal direct sum decomposition

$$\mathscr{C}_G(\tilde{D}) = \bigoplus_B \mathscr{C}_G(\tilde{D}; B)$$

where the sum is taken over the p-blocks of G. So it suffices to show that $\varphi^* - \varphi^G$ is orthogonal to $\mathscr{C}_G(\tilde{D})^0 = \mathscr{C}_G(\tilde{D}; B_0)$ for all $\varphi \in \mathscr{C}_H(D)^0$.

As in the proof of Lemma 30, we can identify the principal blocks of $C_G(a)$ and $C_H(a)$ for each p-element $a \in D$. Hence, following the proof of Theorem 26, we see that $\mathscr{C}_H(D)^0$ and $\mathscr{C}_G(\tilde{D})^0$ have the same dimension and then, since clearly $\ker \sigma = 0$, that σ is surjective. Now, for $\xi \in \mathscr{C}_G(\tilde{D})^0$, there exists $\psi \in \mathscr{C}_H(D)^0$ such that $\xi = \psi^\sigma$. Then

$$\begin{aligned}((\varphi^* - \varphi^G), \xi) &= (\varphi^*, \xi) - (\varphi^G, \xi) \\ &= (\varphi^\sigma, \psi^\sigma) - (\varphi^G, \xi) \\ &= (\varphi, \psi) - (\varphi, \psi^\sigma|_H) \\ &= (\varphi, (\psi - \psi^*|_H))\end{aligned}$$

by Frobenius' reciprocity theorem. However, $\psi^*|_H = \psi$ for $\psi \in \mathscr{C}_H(D)^0$ by definition, and thus $\varphi^* - \varphi^G$ is orthogonal to $\mathscr{C}_G(\tilde{D})^0$ as required.

Now that we have identified σ with the map $\varphi \to (\varphi^G)_0$, the remainder of the theorem is clear by Frobenius reciprocity.

We next turn to the general case of Hypothesis IV where π may consist of more than one prime. Although Reynolds and Robinson have introduced the concept of π-blocks, it is not relevant here. Rather, in order to determine the domain of a candidate for an isometry, we shall find the motivation in the argument of Lemma 30.

Definition. Let X be a group, τ a set of primes and S a union of τ-sections in X. Let $\mathscr{C}_X^\tau(S)^0$ be the subspace of $\mathscr{C}_X(S)$ which consists of those $\theta \in \mathscr{C}_X(S)$ such that $\theta(abc) = \theta(ab)$ whenever a is a τ-element of S, b is a τ-regular element of $C_X(a)$ and $c \in O_{\tau'}(C_X(a))$.

(Notice that, if $\tau \subseteq \pi$, then S may also be a union of π-sections, but $\mathscr{C}_X^\tau(S)^0 \neq \mathscr{C}_X^\pi(S)^0$; thus the definition does depend on τ.)

Let G be a group satisfying Hypothesis IV, and adopt the notation there. As was the case in Lemma 30, if $u \in \tilde{D}$, then u_π is conjugate to an element of D and we may suppose that $u_\pi \in D$ for the purpose of defining class functions on G. We shall define a map

$$*: \mathscr{C}_H^\pi(D)^0 \to \mathscr{C}_G^\pi(\tilde{D})^0$$

which is analogous to σ as follows; this will be *Robinson's isometry*.

Let $\varphi \in \mathscr{C}_H^\pi(D)^0$. Whenever $u \in \tilde{D}$ and $u_\pi \in D$, suppose that $u_{\pi'} = vw$ where v is a π-regular element of $C_H(u_\pi)$ and $w \in O_{\pi'}(C_G(u_\pi))$. Then

$$\varphi^*(u) = \begin{cases} \varphi(u_\pi v) & \text{if } u \in \tilde{D} \text{ and } u_\pi \in D, \\ 0 & \text{if } u \notin \tilde{D}. \end{cases}$$

We see immediately that the definition of $\mathscr{C}_H^\pi(D)^0$ captures the essential features of the proof of Lemma 30 so that * is well-defined and has the desired range; furthermore, the proof of Lemma 31 carries over *verbatim* with π in place of p to show that * is an isometry. (Indeed, we did not take Reynolds' own definition of his isometry but rather a variant which is equivalent, precisely for this reason.) Unfortunately it is not possible to show that * carries generalised characters to generalised characters, though there is no reason to suppose that a proof might not one day be found. However, such an assertion is true for a smaller domain.

Theorem 33. *Assume Hypothesis IV and the definitions above. Suppose that the function* $\theta \in \mathscr{C}_H^\pi(D)^0$ *is a generalised character which is constant on* π*-sections. Then* θ^* *is a generalised character.*

In order to prove Theorem 33, we note that we may replace $\mathbb{C}$ by any subfield which is a splitting field for every subgroup of G; specifically, we shall take the field $K = \mathbb{Q}(\varepsilon)$ where ε is a primitive nth root of unity with n the exponent of G. We first show the following, which does not require θ to take constant values on π-sections.

Lemma 34. *Assume Hypothesis IV and the definitions above. Suppose that the function* $\theta \in \mathscr{C}_H^\pi(D)^0$ *is a generalised character. Then* θ^* *is a rational linear combination of characters of G.*

Proof. For each $\alpha \in \mathrm{Gal}(\mathbb{Q}(\varepsilon))$, there is an integer m coprime to n such that $\varepsilon^\alpha = \varepsilon^m$ and, for each irreducible character χ of G, we may define an irreducible character χ^α by $\chi^\alpha(g) = \chi(g^m)$ by Lemma 2.15. Let $\theta^* = \sum c_i \chi_i$; if we can show that $\theta^*(g^m) = \theta^*(g)^\alpha$ for all $g \in G$, then

$$\sum c_i \chi_i(g^m) = \sum c_i^\alpha \chi_i^\alpha(g) = \sum c_i^\alpha \chi_i(g^m)$$

and hence, since $(m, |G|) = 1$,

$$\sum (c_i^\alpha - c_i)\chi_i = 0.$$

Thus $c_i^\alpha = c_i$ for all i and all α so that $c_i \in \mathbb{Q}$.

We simply compute $\theta^*(g^m)$. We may suppose that either $g \in \tilde{D}$ or $g^m \in \tilde{D}$, for otherwise $\theta^*(g) = \theta^*(g^m) = 0$. (Recall that $\tilde{D}$ is not necessarily closed.) Since $(g^m)_\pi = (g_\pi)^m$, without loss we may suppose that $g_\pi \in D$; also $C_G((g^m)_\pi) = C_G(g_\pi)$. Then, by the choice of the domain of θ, we can effectively compute both $\theta^*(g)$ and $\theta^*(g^m)$ in

$$C_H(g_\pi)O_{\pi'}(C_G(g_\pi))/O_{\pi'}(C_G(g_\pi))$$

or in $C_H(g_\pi)$, so that $\theta^*(g^m) = \theta((g_\pi)^m u^m)$ for any π'-element u of $C_H(g_\pi)$ for which $g = g_\pi uv$ with $v \in O_{\pi'}(C_G(g_\pi))$. But then

$$\theta^*(g^m) = \theta((g_\pi)^m u^m) = \theta(g_\pi u)^\alpha = \theta^*(g)^\alpha$$

since $\theta \in \mathscr{C}\mathscr{h}(H)$.

To complete the proof of Theorem 33, it will be sufficient to show that

θ^* is an algebraic integer combination of characters of G. Since the domain of Robinson's isometry is a certain space of functions over $\mathbb{C}$ and may be restricted to K-linear combinations of characters, we lose nothing by assuming that θ is an algebraic integer combination of characters.

Before proceeding, we establish some consequences of Brauer's characterisation of characters which we shall need.

Lemma 35. *Let L be a group of exponent n, $K = \mathbb{Q}(\varepsilon)$ where ε is a primitive nth root of unity, R the ring of algebraic integers in K and τ a set of primes. Then the following hold.*

(i) *Let x be a τ-element in L. Then the class function which takes values $|L|_\tau$ on $\mathscr{S}_\tau(x)$ and 0 elsewhere lies in $\mathscr{C}h_R(L)$.*
(ii) *Let $\psi \in \mathscr{C}(L, \mathbb{C})$. Suppose that ψ vanishes on all τ-singular elements of L and that $\psi|_N \in \mathscr{C}h_R(N)$ for every nilpotent τ'-subgroup of L. Then $|L|_\tau \psi \in \mathscr{C}h_R(L)$.*
(iii) *If $\psi \in \mathscr{C}h_R(L)$, define a class function ψ' by $\psi'(x) = \psi(x_{\tau'})$. Then $\psi' \in \mathscr{C}h_R(L)$.*

Proof. We shall use Brauer's characterisation of characters in the form of Theorem 5.2′.

(i) It is sufficient to establish the assertion in any nilpotent group, so assume that $L = T \times U$ where T is a τ-group and U is a τ'-group. Then $x \in T$ and, without loss, we may further assume that $U = 1$. By Lemma 5.6, there exists $\xi \in \mathscr{C}h_R(\langle x \rangle)$ such that $\xi(x) = |\langle x \rangle|$ and $\xi(x^i) = 0$ if $x^i \neq x$; then the desired class function on T is $[T:C(x)] \cdot \xi^T$.

(ii) Let $E = T \times N$ be an elementary subgroup of L with T a τ-group and N a τ'-group. Then $\psi|_E$ vanishes outside N and

$$(|L|_\tau \psi)|_E = |L|_\tau \cdot |T|^{-1} (\pi_T \otimes (\psi|_N)) \in \mathscr{C}h_R(E)$$

where π_T is the character of the regular representation of T.

(iii) Since the assertion is clear for irreducible characters of L by an immediate extension of the proof of Lemma 8 (taking X as a τ-subgroup), it follows for elements of $\mathscr{C}h_R(L)$ also.

We now turn to the following hypothesis which, in effect, represents the minimal counterexample to Theorem 33.

Hypothesis V. G is a group, H is a subgroup and D is a subset of H which is a union of π-sections, satisfying the following conditions:

(i) G, H, D satisfy Hypothesis IV;

(ii) if R is the set of algebraic integers in K, then there exists $\theta \in \mathscr{C}\mathscr{h}_R(H) \cap \mathscr{C}_H^\pi(D)^0$ such that $\theta^* \notin \mathscr{C}\mathscr{h}_R(G)$, where $*$ denotes Robinson's isometry from $\mathscr{C}_H^\pi(D)^0$ to $\mathscr{C}_G^\pi(\tilde{D})^0$; and

(iii) subject to (i) and (ii), $|G|$, $|D|$ and $|\pi^*(D)|$ are chosen to be minimal, in that order, where

$$\pi^*(D) = \{p \in \pi \mid p \text{ divides the order of some element of } D\}.$$

The main reduction step in the proof of Theorem 33 (and for other theorems of a similar type) is as follows.

Lemma 36. *Assume Hypothesis V. Then π contains a prime p such that, if $\tau = \pi - \{p\}$,*

(i) *every π-element in D is a p-element, and*

(ii) *G contains a nilpotent τ'-subgroup E such that $\theta^*|_E \notin \mathscr{C}\mathscr{h}_R(E)$.*

Proof. If $|\pi| = 1$, then $\tau = \varnothing$ and the result follows from Brauer's characterisation of characters with E elementary. So we may assume that $|\pi| > 1$.

Let $\pi = \{p_1, \ldots, p_m\}$ and $\tau_i = \pi - \{p_i\}$. Then there are integers a_i such that

$$\theta^* = \sum_i a_i |G|_{\tau_i} \theta^*$$

and hence, for some i,

$$|G|_{\tau_i} \theta^* \notin \mathscr{C}\mathscr{h}_R(G). \tag{7.3}$$

For this i, we write $p = p_i$ and $\tau = \tau_i$. By Lemma 35(i), for any τ-element $x \in H$ (possibly 1), the function which takes values $|G|_\tau$ on $\mathscr{S}_\tau^H(x)$ and vanishes elsewhere lies in $\mathscr{C}\mathscr{h}_R(H)$. Since $\mathscr{S}_\tau^H(x)$ is a union of π-sections of H, the function θ_x defined by

$$\theta_x(h) = \begin{cases} |G|_\tau \theta(h) & \text{if } h \in \mathscr{S}_\tau^H(x), \\ 0 & \text{otherwise,} \end{cases}$$

lies in $\mathscr{C}\mathscr{h}_R(H) \cap \mathscr{C}_H^\pi(D)^0$ so that θ_x^* is defined and $|G|_\tau \theta^* = \sum \theta_x^*$ where the sum is taken over representatives of the conjugacy classes of τ-elements of H. Hence some $\theta_x^* \notin \mathscr{C}\mathscr{h}_R(G)$. Fix such a τ-element x, let $y_1, \ldots, y_r$ be p-elements of $C_H(x)$ such that

$$D \cap \mathscr{S}_\tau^H(x) = \bigcup \mathscr{S}_\pi^H(xy_i),$$

and put

$$D_x = \bigcup \mathscr{S}_\pi^{C_H(x)}(xy_i).$$

We first claim that the triple $(C_G(x), C_H(x), D_x)$ satisfies Hypothesis IV. Any two π-elements of D_x which are conjugate in $C_G(x)$ are conjugate in G and hence in H by hypothesis; therefore they are conjugate in $C_H(x)$

since they have the same τ-part x. Also, if a is a π-element in D_x, then $C_G(a) \subseteq C_G(x)$ and $C_G(a) = C_H(a)O_{\pi'}(C_G(a))$, from which (ii) of Hypothesis IV follows.

Next, let χ be an irreducible character of G such that $(\theta_x^*, \chi) \notin R$. A straightforward calculation shows that

$$\begin{aligned}(\theta_x^*, \chi) &= |G|^{-1} \sum_i \sum_u \theta_x^*(u)\overline{\chi(u)} \\ &= \sum_i |C_G(xy_i)|^{-1} \sum_u \theta_x^*(u)\overline{\chi(u)} \\ &= |C_G(x)|^{-1} \sum_i \sum_u \theta_x^*(u)\overline{\chi(u)} \qquad (7.4)\end{aligned}$$

where the second summations are taken over the elements of the π-sections of xy_i in G, $C_G(xy_i)$ and $C_G(x)$ respectively. Then

$$(\theta_x^*, \chi) = |C_G(x)|^{-1} \sum_u |G|_\tau \theta^*(u)\overline{\chi(u)}$$

where the summation is carried over the elements of

$$\tilde{D}_x = \bigcup_i \mathscr{S}_\pi^{C_G(x)}(xy_i).$$

Let η be the class function on $C_H(x)$ which agrees with $|G|_\tau \theta$ on D_x and which vanishes elsewhere. Then η is the product of $\theta|_{C_H(x)}$ and the class function which takes the value $|G|_\tau$ on $\mathscr{S}_\tau^{C_H(x)}(x)$ and zero elsewhere. The latter is an R-linear combination of characters by Lemma 35(i) and so $\eta \in \mathscr{C}\mathscr{h}_R(C_H(x))$ and $\eta \in \mathscr{C}_{C_H(x)}^\pi(D_x)^0$. Now let ** denote Robinson's isometry from $\mathscr{C}_{C_H(x)}^\pi(D_x)^0$ to $\mathscr{C}_{C_G(x)}^\pi(\tilde{D}_x)^0$. Then, from (7.4), we immediately see that

$$(\theta_x^*, \chi)_G = (\eta^{**}, \chi|_{C_G(x)})_{C_G(x)}.$$

In particular, $\eta^{**} \notin \mathscr{C}\mathscr{h}_R(C_G(x))$. By the minimality of $|G|$ and $|D|$ in Hypothesis V, it follows that $x \in Z(G)$ and $D_x = D$.

We shall next show that $x = 1$, for then D will consist entirely of p-elements, establishing the first part of the lemma. Suppose, then, that $x \neq 1$. Let $D' = x^{-1}D$. It is easy to check that the triple (G, H, D') satisfies Hypothesis IV. Let θ' be the class function on H defined by $\theta'(y) = \theta(xy)$ if $y \in D'$ and $\theta'(h) = 0$ otherwise. Suppose that $\theta^* = \sum_j c_j \chi_j$ where the summation is taken over the irreducible characters of G. By hypothesis, some c_j is not an algebraic integer. If $*'$ denotes the isometry that corresponds to the triple (G, H, D'), since x is a σ-element in $Z(G)$ we then have

$$\theta'^{*'}(y) = \sum_j c_j \omega_j \chi_j(y),$$

where $\omega_j = \chi_j(x)/\chi_j(1)$ is a unit in R, so that $\theta'^{*'} \notin \mathscr{C}\mathscr{h}_R(G)$. On the other hand, $\pi^*(D') = \{p\}$ and $|\pi^*(D')| < |\pi^*(D)|$. This contradicts the minimality assumption in Hypothesis V, and the first part of the lemma is established.

Finally, since θ^* has its support on a union of π-sections whose elements have π-parts which are p-elements, θ^* vanishes on τ-singular elements. Thus, were $\theta^*|_E \in \mathscr{C}\mathscr{h}_R(E)$ for every nilpotent τ'-subgroup of G, then $|G|_\tau \theta^* \in \mathscr{C}\mathscr{h}_R(G)$ by Lemma 35(ii), contrary to (7.3).

At this point, we are close to completing the proof of Theorem 33. Although in what is essentially a minimal counterexample we have shown that the π-elements of D are all p-elements for some prime $p \in \pi$, we should note that π-sections and p-sections are not the same so that we are not yet in a position to employ Reynolds' isometry; we must show that the configuration can be modified to a situation where $\pi = \{p\}$.

Assume the hypothesis of Theorem 33. Suppose that we are in the position of the conclusion of Lemma 36 under Hypothesis V with $\pi \neq \{p\}$ and, thus, $\tau \neq \varnothing$. The function θ vanishes on τ-singular elements of H; we may define a new class function θ' by $\theta'(h) = \theta(h_{\tau'})$ for all $h \in H$ and then $\theta' \in \mathscr{C}\mathscr{h}_R(H)$ by Lemma 35(iii). Furthermore, $h_{\tau'} = h_p h_{\pi'}$ since $\pi^*(D) = \{p\}$, and hence $\theta'(h) = \theta(h_p)$ since θ is constant on π-sections. So θ' is constant on p-sections.

Now let $\hat{D} = \bigcup \mathscr{S}_p^H(x)$, where the union is taken as x ranges over the conjugacy classes of p-elements, and thus of all π-elements, of D. Any two p-elements of $\hat{D}$ which are conjugate in G are conjugate in H since they are π-elements of D. Also, for x a p-element of $\hat{D}$, we have

$$C_G(x) = C_H(x)O_{\pi'}(C_G(x)) = C_H(x)O_{p'}(C_G(x)),$$

so that Hypothesis IV is satisfied by the triple $(G, H, \hat{D})$ with $\pi = \{p\}$. Let $*'$ be Robinson's isometry for this triple. Since θ' is constant on p-sections of H, certainly $\theta' \in \mathscr{C}_H(D)^0$ and so θ' lies in the domain of Reynolds' isometry σ; indeed, $\theta'^{*'} = \theta^\sigma$. Now θ^* and $\theta'^{*'}$ take the same values on τ'-elements of G since $\hat{D}$ consists of the p-sections in H generated by D. $\theta^\sigma \in \mathscr{C}\mathscr{h}_R(G)$ by Theorem 32 and hence $\theta^\sigma|_E \in \mathscr{C}\mathscr{h}_R(G)$ for the τ'-subgroup E of the conclusion of Lemma 36, but then $\theta^*|_E \in \mathscr{C}\mathscr{h}_R(G)$, contrary to Hypothesis V.

This contradiction completes the proof of Theorem 33.

Exercises

1. Investigate the extent to which an analogue of Frobenius reciprocity holds, as in Theorem 3.
2. Derive an analogue of Theorem 7 corresponding to Hypothesis IV and the use of Robinson's isometry.
   [Warning. This is not purely formal: careful counting is necessary!]

Appendix

Although it is a result which does not fall entirely within the goals of this book, nevertheless given that our applications of character theory have been geared very much towards questions that have arisen out of the classification of finite simple groups, no book such as this could be complete without a discussion of Glauberman's Z^*-theorem (1966), a result which has probably had greater application in this area than any other. No proof of Glauberman's theorem is known which does not require a direct application of block theory; however, we have covered enough in Section 6.2 to enable us to give a proof.

In order to state the theorem, we define, for a finite group G, the subgroup $Z^*(G)$ to be the subgroup of G containing $O(G)$ such that $Z^*(G)/O(G) = Z(G/O(G))$.

Theorem. *Let G be a finite group and let S be a Sylow 2-subgroup of G. Suppose that S contains an involution t which is conjugate (in G) to no other element of S. Then $t \in Z^*(G)$; in particular, G is not simple.*

Proof. We shall apply induction to $|G|$, so assume that G is a counterexample of minimal order and that the involution $t \notin Z^*(G)$. It is clear from the hypothesis that t cannot commute with any distinct conjugate t^g. We make extensive use of the Brauer–Fowler lemma (Proposition 2.52). First, we observe that if t^g is a distinct conjugate, then $\langle t, t^g \rangle$ is a dihedral group. If $t^g t$ has even order, then t is not conjugate to t^g in $\langle t, t^g \rangle$ so that t commutes with a conjugate of t^g in $\langle t, t^g \rangle$ and thus in S, which is not so; hence

(1) $t^g t$ must have odd order for all $g \in$ G.

Conversely, this last property clearly implies that t is conjugate to no other element of S, so it is equivalent to our hypothesis; in particular, this implies that our hypothesis holds in G/N whenever $t \notin N$. Thus, to prove the theorem, in particular

(2) we may suppose that $O(G) = 1$.

Suppose that t is the unique involution of S. Then it is well known that

S must be either cyclic or a generalised quaternion group. In the former case, G has a normal 2-complement and the conclusion holds, while the proof of Brauer–Suzuki theorem actually establishes precisely the desired conclusion in the latter. So we may assume that

(3) G contains a conjugacy class of involutions not containing t.

Furthermore, if s is an involution of G which is not conjugate to t, then $\langle s,t\rangle$ has order divisible by 4 but not 8 since, otherwise, $\langle s,t\rangle$ would contain a conjugate t^x of t different from t such that $\langle t,t^x\rangle$ is a 2-group, again by Proposition 2.52. Thus,

(4) if s is an involution of G which is not conjugate to t, then st has twice odd order.

Further, were $s\in Z(G)$, then $O(G/\langle s\rangle)=1$ by (2) and, by the remark preceding (2), we would have $t\langle s\rangle\in Z(G/\langle s\rangle)$ from which it would follow that $t\in Z(G)$. Hence

(5) if z is any involution of G, then $C_G(z)\neq G$.

In particular, this will allow us to use our inductive hypothesis;

(6) $t\in Z^*(C_G(z))$ for any involution $z\in C_G(t)$.

With these reductions, we shall now proceed more formally. We shall apply block theoretic methods with $p=2$.

Lemma 1. *Let s be an involution in $S-\{t\}$. If s' and t' are any conjugates in G of s and t respectively and χ is an irreducible character in $B_0(G)$, then $\chi(t's')=\chi(ts)$.*

Proof. Without loss, we may suppose that $t'=t$. If $s'\in C(t)$, then s' is conjugate to s in $C(t)$ by a weak closure argument and $\psi(ts')=\psi(ts)$ for any irreducible character ψ of $C(t)$; then $\chi(ts)=\chi(ts')$ for any character of G.

Suppose, then, that $s'\notin C(t)$. We shall construct a conjugate s'' of s such that $s''\in C(t)$ and $\chi(ts'')=\chi(ts')$. Let $u=ts'$ and let z be the involution in $\langle u\rangle$. If $H=C_G(z)$, then $t\in Z^*(H)$ by induction, and $u^2=[t,s']\in O(H)$. If u has order $2n$, then $u^{n+1}=zu\in O(H)$ and, if φ is any member of a basic set for $B_0(H)$, then $\varphi(zu)=\varphi(1)$ by Theorem 6.12. Thus, applying Theorem 6.15, we see that

$$\chi(ts')=\chi(u)=\chi(z\cdot zu)=\chi(z).$$

Now, putting $s''=tz$, s'' is not conjugate to t in $\langle t,s'\rangle$ and hence is conjugate to s'; also $ts''=z$ so that $\chi(ts'')=\chi(ts')$, while $s''\in C_G(t)$.

Lemma 2. *Let s be an involution in S different from t and let χ be a nonprincipal character in $B_0(G)$. If $\chi(s) \neq 0$, then $\chi(t) = -\chi(1)$.*

Proof. Suppose that the conjugacy classes in G are ordered so that $t \in \mathscr{C}_1$ and $s \in \mathscr{C}_2$. Then, computing in the class algebra with the standard notation of Section 2.1, we have nonnegative integers a_j such that

$$C_1 \cdot C_2 = \sum_j a_j C_j.$$

Computation of the central character corresponding to χ leads to an equation

$$\frac{h_1 \chi(t)}{\chi(1)} \cdot \frac{h_2 \chi(s)}{\chi(1)} = \sum_j a_j \frac{h_j \chi(g_j)}{\chi(1)}.$$

If $a_j \neq 0$, then $\chi(g_j) = \chi(ts)$ by Lemma 1. Hence

$$h_1 h_2 \chi(t)\chi(s) = \chi(1)\chi(ts) \sum_j a_j h_j.$$

However, by counting elements, we see that $h_1 h_2 = \sum_j a_j h_j$; hence

$$\chi(t)\chi(s) = \chi(1)\chi(ts).$$

This did not require that $\chi(s) \neq 0$; hence we may replace s by ts to obtain a further equation

$$\chi(t)\chi(ts) = \chi(1)\chi(s).$$

Now, since $\chi(s) \neq 0$, this implies that

$$\frac{\chi(1)}{\chi(t)} = \frac{\chi(t)}{\chi(1)}$$

so that $\chi(t) = \pm \chi(1)$. However, if $\chi(t) = \chi(1)$, then $t \in \ker \chi$ and it follows by induction that $t \in Z^*(\ker \chi)$ and then that $t \in Z(G)$, a contradiction. Hence $\chi(t) = -\chi(1)$.

We may now easily complete the proof of Glauberman's theorem. By Exercise 3 of Section 6.2,

$$\sum_\chi \chi(t)\chi(s) = 0$$

and

$$\sum_\chi \chi(1)\chi(s) = 0$$

where s is an involution not conjugate to t and the summations are carried over $B_0(G)$. Thus

$$\sum_\chi (\chi(t) + \chi(1))\chi(s) = 0.$$

By Lemma 2,

$$2 = 2 + \sum{}'(\chi(t) + \chi(1))\chi(s) = 0$$

where the summation is taken over the nonprincipal characters in $B_0(G)$, which is the final contradiction.

References

Books

[A] J. L. Alperin, *Local representation theory*, Cambridge University Press, Cambridge, 1986.

[C] R. Carter, *Finite groups of Lie type*, Wiley, Chichester, 1985.

[CR I] C. W. Curtis and I. Reiner, *Methods of representation theory* I, Wiley, New York, 1981.

[CR II] C. W. Curtis and I. Reiner, *Methods of representation theory* II, Wiley, New York, 1987.

[F] W. Feit, *Characters of finite groups*, Benjamin, New York, 1967.

[F*] W. Feit, *The representation theory of finite groups*, North-Holland, Amsterdam, 1982.

[G] D. Gorenstein, *Finite groups*, Harper and Row, New York, 1968.

[H] B. Huppert, *Endliche Gruppen* I, Springer-Verlag, Berlin, 1967.

[HB II] B. Huppert and N. Blackburn, *Finite groups* II, Springer-Verlag, Berlin, 1982.

[HB III] B. Huppert and N. Blackburn, *Finite groups* III, Springer-Verlag, Berlin, 1982.

[JK] G. D. James and A. Kerber, *The representation theory of the symmetric group*, Cambridge University Press, Cambridge, 1984.

[S I] M. Suzuki, *Group Theory* I, Springer-Verlag, Berlin, 1982.

[S II] M. Suzuki, *Group Theory* II, Springer-Verlag, Berlin, 1986.

Articles

G. V. Belyi (1980) On Galois extensions of a maximal cyclotomic field, *Math. USSR Izv., Amer. Math. Soc. Translations* **14**, 247–56.

H. Bender (1974) The Brauer–Suzuki–Wall theorem, *Illinois J. Math.* **18**, 229–35.

R. Brauer (1942) On groups whose order contains a prime number to the first power I, *Amer. J. Math.* **64**, 401–20.

(1947) Applications of induced characters, *Amer. J. Math.* **69**, 709–16.

(1953) A characterization of the characters of groups of finite order, *Ann. of Math.* (2) **57**, 357–77.

(1964; I) Some applications of the theory of blocks of characters of finite groups I, *J. Algebra* **1**, 152–67.

(1964; II) Some applications of the theory of blocks of characters of finite groups II, *J.* Algebra **1**, 307–34.

R. Brauer and K. A. Fowler (1955) On groups of even order, *Ann. of Math.* (2) **62**, 565–83.

R. Brauer and M. Suzuki (1959) On finite groups of even order whose 2-Sylow subgroup is a quaternion group, *Proc. Nat. Acad. Sci. U.S.A.* **45**, 1757–9.

R. Brauer, M. Suzuki and G. E. Wall (1958) A characterization of the one-dimensional unimodular groups over finite fields, *Illinois J. Math.* **2**, 718–45.

R. Brauer and J. Tate (1955) On the characters of finite groups, *Ann. of Math.* (2) **62**, 1–7.

W. Burnside (1900) On a class of groups of finite order, *Transactions of the Cambridge Philos. Soc.* **18**, 269–76.

A. R. Camina and M. J. Collins (1974) Finite groups admitting automorphisms with prescribed fixed points, *Proc. London Math. Soc.* (3) **28**, 45–66.

A. H. Clifford (1937) Representations induced in an invariant subgroup, *Ann. of Math.* (2) **38**, 533–50.

E. Cline (1972) Stable Clifford theory, *J. Algebra* **22**, 350–64.

M. J. Collins (1988) Characters of finite groups having a self-normalising cyclic subgroup, *J. Algebra* **119**, 282–97.

E. C. Dade (1964) Lifting group characters, *Ann. of Math.* (2) **79**, 590–6.

(1970) Compounding Clifford's theory, *Ann. of Math.* (2) **91**, 236–90.

(1980) Group-graded rings, *Math. Z.* **174**, 241–62.

W. Feit (1960) On a class of doubly transitive permutation groups, *Illinois J. Math.* **4**, 170–86.

(1984) Rigidity and Galois groups, *Proceedings of the Rutgers group theory year, 1983–1984* (ed. M. Aschbacher, D. Gorenstein, R. Lyons, M. O'Nan, C. Sims & W. Feit), Cambridge University Press, pp. 283–7.

W. Feit, M. Hall Jr. and J. G. Thompson (1960) Finite groups in which the centralizer of any non-identity element is nilpotent, *Math. Z.* **74**, 1–17.

W. Feit and J. G. Thompson (1962) Finite groups which contain a self-centralizing subgroup of order 3, *Nagoya J. Math.* **21**, 185–97.

(1963) Solvability of groups of odd order, *Pacific J. Math.* **13**, 775–1029.

G. Glauberman (1966) Central elements in core-free groups, *J. Algebra* **4**, 403–20.

(1969) On a class of doubly transitive permutation groups, *Illinois J. Math.* **13**, 394–9.

(1974) On groups with a quaternion Sylow 2-subgroup, *Illinois J. Math.* **18**, 60–5.

D. Goldschmidt (1970) A group theoretic proof of the $p^a q^b$ theorem for odd primes, *Math. Z.* **113**, 373–5.

D. Goldschmidt and I. M. Isaacs (1975) Schur indices in finite groups, *J.* Algebra **33**, 191–9.

D. Gorenstein and J. H. Walter (1962) On finite groups with dihedral Sylow 2-subgroups, *Illinois J. Math.* **6**, 553–93.

J. A. Green (1955) On the converse to a theorem of R. Brauer, *Proc. Cambridge Philos. Soc.* **51**, 237–9.

P. Hall and G. Higman (1956) On the p-length of p-soluble groups and reduction theorems for Burnside's problem, *Proc. London Math. Soc.* (3) **6**, 1–42.

K. Harada (1967) A characterization of the groups $LF(2, q)$, *Illinois J. Math.* **11**, 647–59.

D. G. Higman (1953) Focal series in finite groups, *Canad. J. Math.* **5**, 477–97.

G. Higman (1968) Odd characterisations of simple groups, Univ. of Michigan (mimeographed notes).

D. C. Hunt (1986) Rational rigidity and the sporadic groups, *J. Algebra* **99**, 577–92.

N. Itô (1962) On a class of doubly transitive permutation groups, *Illinois J. Math.* **6**, 341–52.

G. W. Mackey (1951) On induced representations of groups, *Amer. J. Math.* **73**, 576–92.

H. Matsuyama (1973) Solvability of groups of order $2^a p^b$, *Osaka J. Math.* **10**, 375–8.

M. Osima (1955) Notes on blocks of groups characters, *Math. J. Okayama Univ.* **4**, 175–88.

W. F. Reynolds (1967) Sections, isometries, and generalized group characters, *J. Algebra* **7**, 394–405.

(1968) Isometries and principal blocks of group characters, *Math. Z.* **107**, 264–70.

G. R. Robinson (1985) Blocks, isometries, and sets of primes, *Proc. London Math. Soc.* (3) **51**, 432–48.

D. A. Sibley (1975) Finite linear groups with a strongly self-centralizing Sylow subgroup, *J. Algebra* **36**, 319–32.

(1976) Coherence in finite groups containing a Frobenius section, *Illinois J. Math.* **20**, 434–42.

S. D. Smith and A. P. Tyrer (1973) On finite groups with a certain Sylow normalizer II, *J. Algebra* **26**, 366–7.

W. B. Stewart (1967) Problems in group theory, D. Phil. thesis, Oxford.

M. Suzuki (1955) On finite groups with cyclic Sylow subgroups for all odd primes, *Amer. J. Math.* **77**, 657–91.

(1957) The nonexistence of a certain type of simple group of odd order, *Proc. Amer. Math. Soc.* **8**, 686–95.

(1959) Applications of group characters, *Proc. Sympos. Pure Math.* Vol. 1, Amer, Math. Soc., Providence, R.I., pp. 88–99.

(1962a) Applications of group characters, *Proc. Sympos. Pure Math.*, Vol. VI, Amer. Math. Soc., Providence, R.I., pp. 101–5.

(1962b) On a class of doubly transitive groups, *Ann. of Math.* (2) **75**, 105–45.

(1963) On the existence of a Hall normal complement, *J. Math. Soc. Japan* **15**, 387–91.

(1964) Finite groups of even order in which Sylow 2-groups are independent, *Ann. of Math.* (2) **80**, 58–77.

J. G. Thompson (1959) Finite groups with fixed-point-free automorphisms of prime order, *Proc. Nat. Acad. Sci. U.S.A.* **45**, 578–81.

(1960) Normal p-complements for finite groups, *Math. Z.* **72**, 332–54.

(1964) Normal p-complements for finite groups, *J. Algebra* **1**, 43–6.

(1984) Some finite groups which occur as Gal L/K, where $K \subseteq \mathbb{Q}(\mu_n)$, *J. Algebra* **89**, 437–99.

W. J. Wong (1966) Exceptional character theory and the theory of blocks, *Math. Z.* **91**, 363–79.

T. Yoshida (1978) Character-theoretic transfer, *J. Algebra* **52**, 1–38.

H. Zassenhaus (1936) Kennzeichnung endlicher linearer Gruppen als Permutationsgruppen, *Abh. Math. Sem. Univ. Hamburg* **11**, 17–44.

Index of notation

The following is a partial list, giving standard notation which is not necessarily defined in the text (especially that for groups) and notation which is used other than in the section in which it is defined or where it is similar to other notation. Page numbers refer to a definition or first appearance.

Let G be a group, H a subgroup, p a prime and P a p-subgroup. Other lower case letters denote elements. π is a set of primes and π' the complementary set.

$x^g = g^{-1}xg$	conjugate
$[x, y] = x^{-1}y^{-1}xy$	commutator
$g \sim_G g'$	g is conjugate (in G) to g'
$C(\ )$, $N(\ )$	centraliser, normaliser
$Z(G)$	centre of G
$H \trianglelefteq G$	H is a normal subgroup of G
G'	derived group
$F(G)$	Fitting subgroup (maximal nilpotent normal subgroup)
$G^{\#}$	set of nonidentity elements of G
x^G	set of conjugates of x (and similarly for subsets of G)

In the following, K will be a field, L an extension of K, V a vector space over K; ρ will denote a representation (either of G or of a subgroup). A will be a K-algebra and M, N will be A-modules. (Often, in the first use, we shall take $K = \mathbb{C}$, and A a group algebra.)

General index